Advanced Structured Materials

Volume 62

More information about this series at http://www.springer.com/series/8611

Ashutosh Kumar Shukla
Editor

EMR/ESR/EPR Spectroscopy for Characterization of Nanomaterials

Springer

Editor
Ashutosh Kumar Shukla
Department of Physics
Ewing Christian College
Allahabad
India

ISSN 1869-8433 ISSN 1869-8441 (electronic)
Advanced Structured Materials
ISBN 978-81-322-3653-5 ISBN 978-81-322-3655-9 (eBook)
DOI 10.1007/978-81-322-3655-9

Library of Congress Control Number: 2016950380

Printed on acid-free paper

This Springer imprint is published by Springer Nature
The registered company is Springer (India) Pvt. Ltd.
The registered company address is: 7th Floor, Vijaya Building, 17 Barakhamba Road, New Delhi 110 001, India

To my parents

Preface

Nanotechnology has opened up a new area of research in different fields. Advantages are as many as to overcome disadvantages. The present collection intends to describe the applications of electron paramagnetic resonance (EPR) spectroscopy in the area of nanomaterial characterization. I have used all the three names in the title of the book for the technique, viz, electron magnetic resonance (EMR), electron paramagnetic resonance (EPR) and electron spin resonance (ESR). Logically EMR includes the terms EPR and ESR. Most of the available books on EPR generally involve mathematics and quantum mechanics. It poses a hurdle before the scientists from other disciplines. I have tried to avoid these mathematical details to make this collection more meaningful for readers from different disciplines.

This book contains two parts. First part deals with nanomaterials and their classification. Second part deals with EMR characterization of nanomaterials. Chapter "Nanomaterials and their Classification" by Cristina Buzea and Ivan Pacheco is an excellent attempt to acquaint the reader with the nanomaterials and highlights their physico-chemical properties. Chapter "A Brief Manifestation of Nanotechnology" by Sharda Sundaram Sanjay and Avinash C. Pandey complements Chapter "Nanomaterials and their Classification" and includes applications based classification. A section on magnetization dynamics of nanomaterials which is covered in this chapter provides a link to the second part on EMR characterization of nanomaterials. Chapter "An Overview on Advances in the NanoCarriers Drug Delivery Systems" by Anjana Pandey gives an overview of the recent trends of nanocarriers drug delivery systems which have attracted the attention of the scientific community due to several advantages over the conventional drug delivery systems.

Applications of EMR spectroscopy covered in second part of the book deal basically with two different categories of nanomaterials—metallic nanoparticles and metal oxide nanoparticles. Chapter "EMR of Metallic Nanoparticles" by Siavash Iravani deals with EMR of metallic nanoparticles. Chapter "Electron Spin Resonance Applied to Nanosized-Doped Oxides" by Cesare Oliva and Marco Scavini covers the ESR of nanosized doped oxides. Chapter "EPR Studies of

Cerium Dioxide Nanoparticles" by me and my co-author Rafail Rakhmatullin deals with recent EPR applications in the characterization of ceria nanoparticles. Most of this chapter is based on the research work of Rafail Rakhmatullin and co-workers. Chapter "Synthesis and Characterization of Undoped and Doped (Mn, Cu, Co) ZnO Nanoparticles: An EPR Study" by Seyda Colak and Cangul Akturk covers EPR of zinc oxide nanoparticles.

I am grateful to Prof. Ram Kripal, Head, Department of Physics, University of Allahabad who has introduced me to the language of EMR research. I sincerely thank Swati Meherishi, Publishing Editor, Springer for giving me an opportunity to present this book to the readers. My special thanks to Dr. Natalia Noginova (Norfolk State University, Norfolk, VA), Dr. Emre Erdem (University of Freiburg, Germany), Dr. Bernadeta Dobosz (Adam Mickiewicz University, Poland), and Dr. Thomas Yeager (Victoria University, Melbourne, Australia) for taking their time to go through the chapters and provide their valuable reviews. I would also like to thank the authors for their contributions.

My sincere thanks to Prof. Raja Ram Yadav, Department of Physics, University of Allahabad, Dr. M. Massey, Principal, Ewing Christian College, Allahabad and my colleagues Dr. R.C. Mital, Dr. Anjani Kumar Singh, Dr. Mrs. Kusum Lata Pandey, and Dr. Anil Kumar Singh for their constant encouraging remarks and suggestions during the development of this book.

It is difficult to express my gratitude in words to my brother Dr. Arun K. Shukla, Department of Biological Sciences and Bioengineering, Indian Institute of Technology, Kanpur, who has always been there to help me. My special thanks are also due to my wife Dr. Neelam Shukla.

In spite of our efforts, errors and omissions might have crept in. I request esteemed readers to bring these mistakes to our notice.

Allahabad, India
December 2015

Ashutosh Kumar Shukla

Contents

About the Editor

Dr. Ashutosh Kumar Shukla has obtained his B.Sc., M.Sc., and D.Phil. degrees from the University of Allahabad.

His talks on science have been regularly broadcasted on the radio. He has been working to popularize science among the masses through his published science articles on issues of general interest. He has received many scholarships and fellowships including National Scholarship, Scholarship of Ministry of Higher Education, Government of UP, Research Fellowship of Council of Science and Technology, Lucknow, UP, Research Fellowship of Council of Scientific and Industrial Research, New Delhi and Indian National Science Academy-Bilateral Exchange Fellowship.

During his doctoral work he focused on electron spin resonance spectroscopy and optical absorption spectroscopy to investigate transition ion doped single crystals. In addition to his exposure to the available CW-ESR spectrometers at leading institutes of his own country like National Physical Laboratory, New Delhi, Indian Institute of Technology, Bombay and Bhabha Atomic Research Center, Bombay, he has hands-on experience with modern CW and pulsed ESR spectrometers at several leading international laboratories including the University of Dundee, Scotland, University of St. Andrew's, Scotland and Kazan State University, Kazan. ESR characterization of rare earth manganites and nanomaterials are his current interests.

Part I
Nanomaterials and their Classification

Nanomaterials and their Classification

Cristina Buzea and Ivan Pacheco

Abstract In this chapter, we present a general classification of nanomaterials based on their dimensionality, shape, and composition. According to their dimensionality, nanomaterials include nanoparticles, nanotubes, and nanofilms. Nanomaterials can be made of single elements, such as metals or carbon, or multiple elements, such as metal oxides or composites. We review the most used types of nanomaterials up to date showing examples of their morphologies. The physicochemical properties of a material in nanoform can be very different from its bulk counterpart, depending not only on the type of materials, but on its size, shape, and functionalization. We discuss the most important physicochemical properties of nanomaterials, among which are morphology, dispersability, crystalline phase, melting temperature, and magnetic properties.

Keywords Nanomaterials classification · Nanoparticles · Nanomaterials physicochemical properties · Magnetic properties of nanoparticles · Toxicity of nanoparticles · Melting temperature of nanoparticles

C. Buzea (✉)
IIPB Medicine Corporation, Owen Sound, ON N4K 6S5, Canada
e-mail: cristinabuzea@mdcorporation.ca

I. Pacheco
Department of Pathology, Grey Bruce Health Services, 1800 8th St East, Owen Sound, ON N4K 6M9, Canada
e-mail: ipacheco@gbhs.on.ca; iblandin@uwo.ca

I. Pacheco
Department of Pathology and Laboratory Medicine, Schulich School of Medicine & Dentistry, Western University, London, ON N6A 5C1, Canada

A.K. Shukla (ed.), *EMR/ESR/EPR Spectroscopy for Characterization of Nanomaterials*, Advanced Structured Materials 62,
DOI 10.1007/978-81-322-3655-9_1

1 Nanomaterials Definition and Classification

1.1 Nanomaterials Definition

A **nanomaterial** can be defined as a material with external dimensions in the nanoscale or with internal structure or surface structure in the nanoscale. According to this definition, most of the materials around us would qualify as nanomaterials, as their internal structure is modulated at the nanoscale (See for example Buzea et al. [21]).

A **nanoparticle** is defined as a particle with at least one external dimension in the nanoscale.

The term **nanoscale** should be defined as a size range from about 1 nm to 1000 nm. The upper limit overlaps with the definition of microparticles (whose lower size limit is about 1 micron). While the lower limit of 1 nm is commonly accepted, an upper limit of only 100 nm has been adopted by some authors, however this limit has no scientific evidence to support the choice of this value, and the use of a single limit might be too limiting for the classification of nanomaterials.

A differential approach for the term of "nanoparticle" was proposed in 2011 by the European Commission [113]. The number size distribution should be taken into account using the mean size and the standard deviation of the size to refine the definition.

Nanoparticles consist usually of agglomeration of particles with sizes in a particular distribution, some larger and some smaller than 100 nm limit. The European commission suggests that a nanomaterial should consist of 50 % or more of particles having a size between 1 and 100 nm [113].

The European Union agencies suggest that the following definition should apply to the term "nanomaterial" [113]:

> 'Nanomaterial' means a natural, incidental or manufactured material containing particles, in an unbound state or as an aggregate or as an agglomerate and where, for 50 % or more of the particles in the number size distribution, one or more external dimensions is in the size range 1–100 nm. In specific cases and where warranted by concerns for the environment, health, safety or competitiveness the number size distribution threshold of 50 % may be replaced by a threshold between 1 and 50 %. By derogation from the above statement, fullerenes, graphene flakes and single wall carbon nanotubes with one or more external dimensions below 1 nm should be considered as nanomaterials.
>
> 'Particle', 'agglomerate' and 'aggregate' are defined as follows:
>
> (a) 'particle' means a minute piece of matter with defined physical boundaries;
> (b) 'agglomerate' means a collection of weakly bound particles or aggregates where the resulting external surface area is similar to the sum of the surface areas of the individual components;
> (c) 'aggregate' means a particle comprising of strongly bound or fused particles.

"Nano" should be considered a different state of aggregation of matter in addition to the solid, liquid, gas, and plasma states. The reason for this is because nanoparticles of a given material have quite different physical (optical, magnetic)

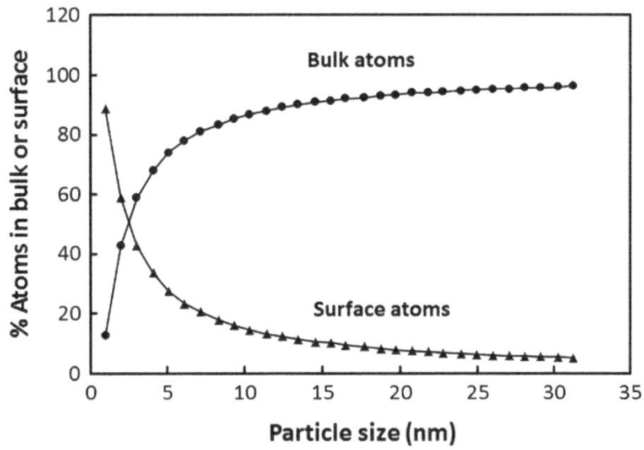

Fig. 1 Calculated surface–to-bulk ratios for solid metal particles versus size. Adapted from Klabunde et al. [73]

chemical, and mechanical properties than the same material in bulk form [91], [77]. Two main factors contribute to this: surface effects and quantum effects that appear with decrease in size. Surface effects lead to a smooth scaling of physical properties due to an increased fraction of the atoms at the particle surface compared to the interior (Fig. 1) [73]. Surface effects result in increased chemical reactivity, reduced melting point of nanoparticles compared to larger particles or bulk material. Quantum-size effects involve the confinement of electrons within a very small nanoparticle, or quantum dot which manifest in a quantized energy spectrum, resulting in the appearance of magnetic moments, particularly in nanoparticles made of nonmagnetic materials in bulk form such as gold, platinum, and palladium.

A nanoparticle can be thought as an aggregate composed of a relatively small number of atoms. The smaller nanoparticles, sometimes called clusters, consist of only a few atoms. The larger nanoparticles can have more than 10^5 atoms. Nanoparticles are objects with intermediate size between the domain of atoms, which must be described using quantum mechanics, and the bulk materials that can be described by classical electrodynamics and solid-state physics [69]. Therefore, nanoparticles are objects at mesoscopic scale between the microscopic and macroscopic worlds. (Mesoscopic scale means pertaining to a scale between microscopic and macroscopic, or an intermediate scale.) Because of this some nanoparticles exhibit a number of new phenomena that we will try to capture in this chapter.

1.2 Nanomaterials Classifications

1.2.1 Dimensionality

From the point of view of their dimensionality, nanomaterials can be classified as nanomaterials with one, two, and three dimensions within the nanoscale, see Fig. 2 [21]. Materials with one dimension in the nanoscale are also called very thin films or surface coatings attached on a substrate usually made from a different material. Nanomaterials with two dimensions in the nanoscale are either nanoparticles attached onto a substrate, porous thin films with pores in the nanoscale, or free long aspect ratio nanoparticles, wires, or tubes. Finally, nanomaterials with three dimensions within the nanoscale can be fixed small nanostructures on a substrate, membranes with nanopores on a substrate, or nanoparticles. In this chapter one will focus mainly on nanoparticles.

Examples of nanoparticles of various nanomaterials are illustrated in Figs. 3, 4, and 5. Figure 3 shows transmission electron microscopy (TEM) images of various metal oxide nanoparticles: Al, Sb, Bi, Co, Cr, Fe, In, La, Mn, Ni, Si, Sn, Ti, W, V, Y, Zn, and Zr [38]. Each material shows a specific aggregation pattern of the nanoparticles, with aggregation depending on their surface chemistry, some materials have crystalline nanoparticles, while others are polycrystalline.

Figure 4 shows TEM images of Ag nanoparticles of various size ranges together with plots of nanoparticle distribution [2]. In general, the particle distribution follows a Gaussian bell.

Figure 5 illustrates microscopy images of different shaped silica nanoparticles [57]. The same material can be shaped in different forms, spherical, oval, or nanorod forms.

When we think of long aspect ratio nanoparticles, we think carbon nanotubes, however, a wide range of materials can be synthesized as nanoparticles with long aspect ratio. Figure 6 illustrates some examples of long aspect ratio nanoparticles:

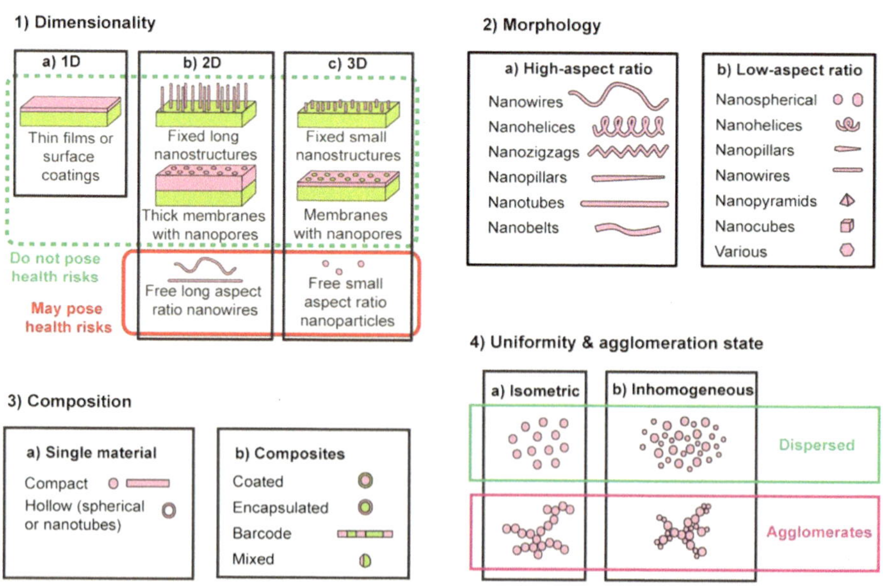

Fig. 2 Classification of nanostructured materials according to their dimensionality, morphology, composition, uniformity, and agglomeration state [21]

Fig. 3 Transmission electron microscopy images of various metal oxide nanoparticles (MeO_x). **a** aluminum, **b** antimony, **c** bismuth, **d** cobalt II, **e** chromium, **f** iron, **g** indium, **h** lanthanum, **i** manganese, **j** nickel, **k** silicon, **l** tin, **m** titanium, **n** tungsten, **o** vanadium, **p** yttrium, **q** zinc, **r** zirconium. Reprinted with permission from: "Towards understanding mechanisms governing cytotoxicity of metal oxides nanoparticles: Hints from nano-QSAR studies," Gajewicz et al. Nanotoxicology, 2015 Taylor & Francis Ltd [38]

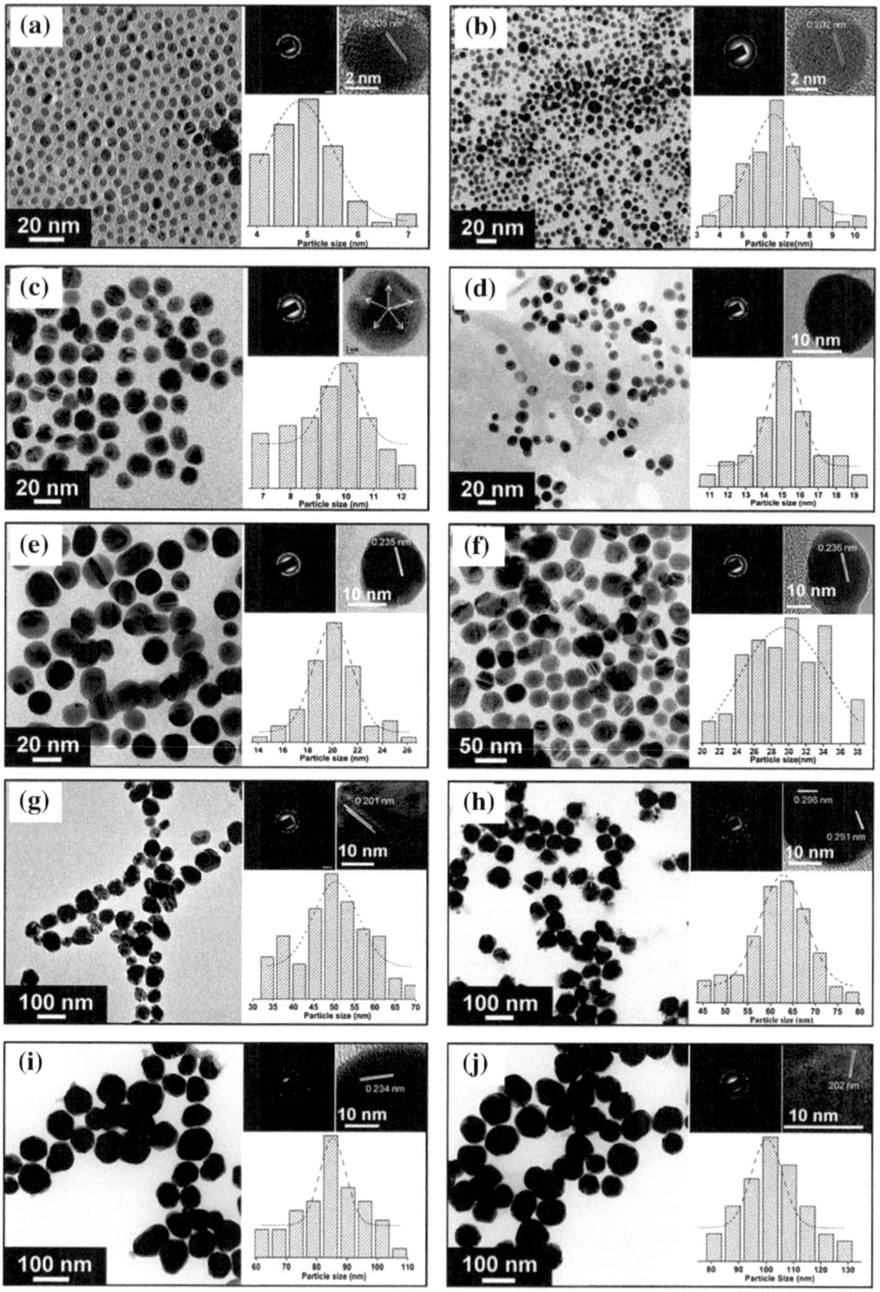
(a)
20 nm
2 nm
Particle size (nm)
(b)
20 nm
2 nm
Particle size(nm)
(c)
20 nm
Particle size (nm)
(d)
20 nm
10 nm
Particle size (nm)
(e)
20 nm
0.235 nm
10 nm
Particle size (nm)
(f)
50 nm
0.236 nm
10 nm
Particle size(nm)
(g)
100 nm
0.201 nm
10 nm
Particle size (nm)
(h)
100 nm
0.296 nm
0.291 nm
10 nm
Particle size (nm)
(i)
100 nm
0.234 nm
10 nm
Particle size (nm)
(j)
100 nm
202 nm
10 nm
Particle Size (nm)

◀ **Fig. 4** Field emission gun-transmission electron microscopy (FEG-TEM) images of silver nanoparticles of various size ranges **a** 5 ± 0.7 nm; **b** 7 ± 1.3 nm; **c** 10 ± 2.0 nm; **d** 15 ± 2.3 nm; **e** 20 ± 2.5 nm; **f** 30 ± 5.1 nm; **g** 50 ± 7.1; **h** 63 ± 7.6; **i** 85 ± 8.2; **j** 100 ± 11.2 nm. For each image, the corresponding high-resolution (HRTEM) image and lattice fringes (d-spacing) are shown. The histograms in the right lower side indicate the particle size distribution. Reproduced from Agnihotri and Mukherji [2]

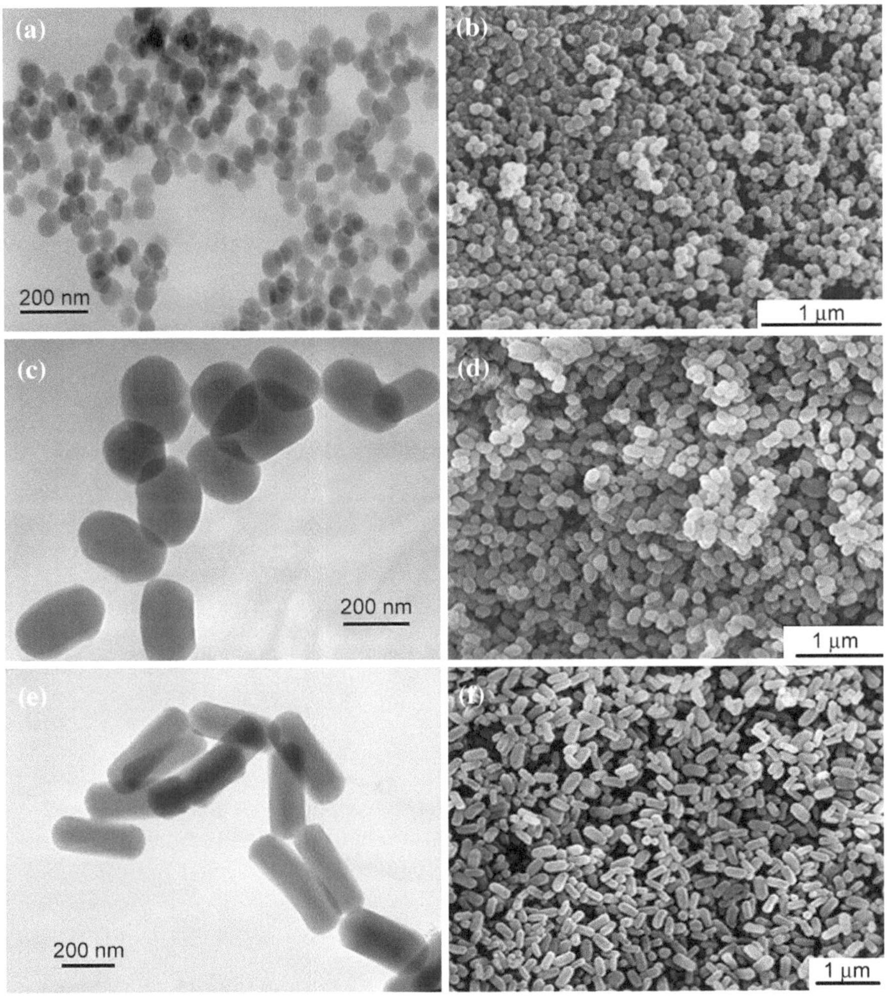

Fig. 5 **a**, **c**, **e** Transmission electron microscopy and **b**, **d**, **f** scanning electron microscopy images of different shaped silica nanoparticles; **a**, **b** sphere-shaped nanoparticles, **c**, **d** rod-shaped nanoparticles, **e**, **f** long rod-shaped nanoparticles. Reprinted from Biomaterials, Volume 31, Issue 3, Huang X et al., The effect of the shape of mesoporous silica nanoparticles on cellular uptake and cell function, pp. 438–448, Copyright (2010), with permission from Elsevier [57]

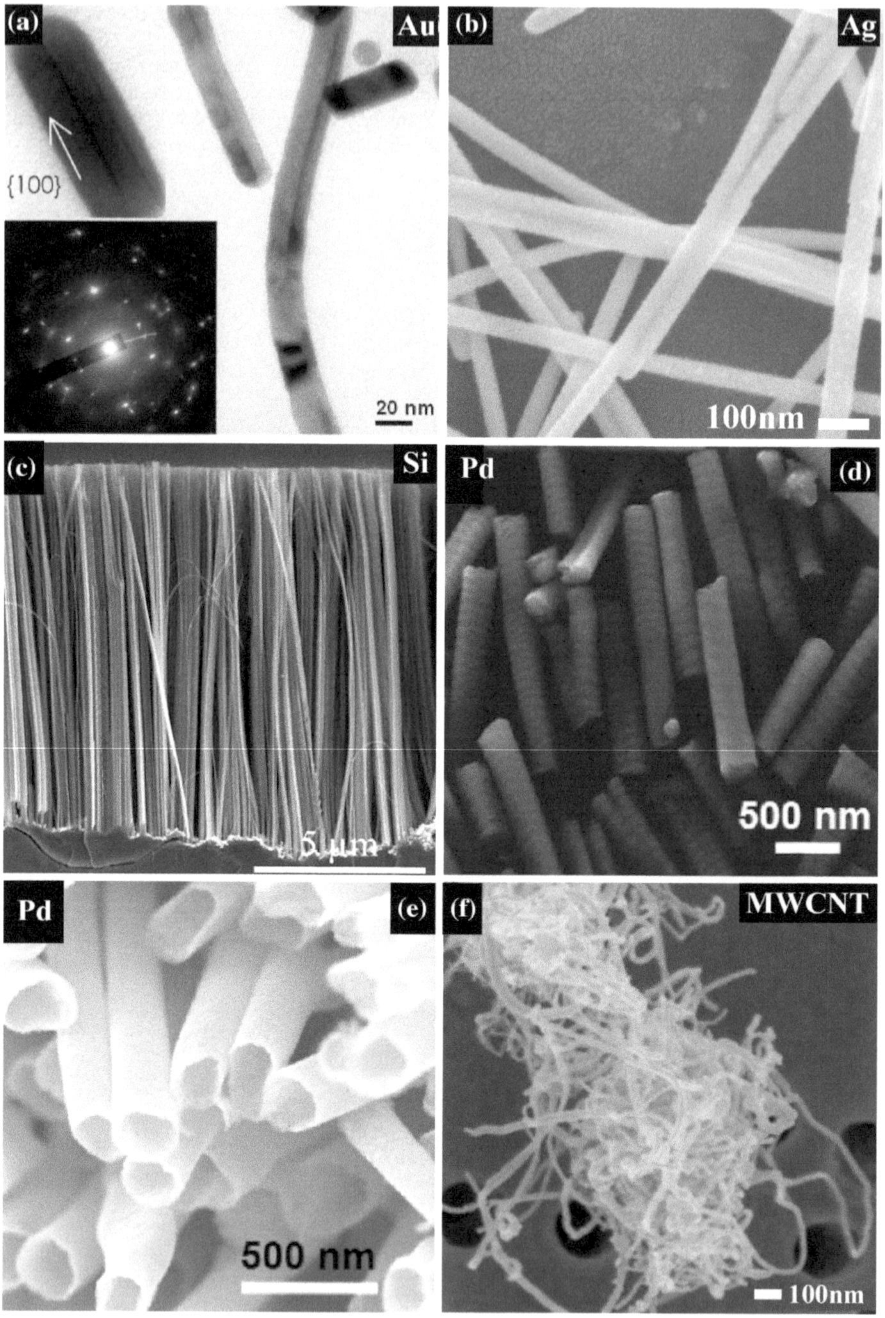
(a)
Au
{100}
20 nm
(b)
Ag
100nm
(c)
Si
5 μm
Pd
(d)
500 nm
Pd
(e)
500 nm
(f)
MWCNT
100nm

◀ **Fig. 6** **a** TEM (corresponding electron diffraction pattern inset, lower) and **b–f** SEM images of long aspect ratio nanomaterials. Nanowires of **a** Au, **b** Ag, **c** Si attached on a substrate, **d** Pd, and **e** nanotubes of Pd, and **f** multi-walled carbon nanotubes. Figure a reproduced from Das et al. [33] with permission from John Wiley and Sons. Figure b reprinted from Journal of Hazardous Materials, Volume 177, B. Cheng et al., Preparation and enhanced photocatalytic activity of Ag@TiO2 core–shell nanocomposite nanowires, pp. 971–977, Copyright (2010), with permission from Elsevier [25]. Figure c reprinted by permission from Macmillan Publishers Ltd: Scientific Reports [56] copyright (2014). Figures d and e reprinted with permission from nanoletters, Volume 11, issue 9, Liu et al. Wet-Chemical Synthesis of Palladium Nanosprings, pp. 3979–3982, Copyright (2011) American Chemical Society [87]. Figure f reproduced from Ma-Hock et al. [93]

nanowires of Au, Ag, and Si, Pd, nanotubes, and multi-walled carbon nanotubes (MWCNT) [25, 33, 56, 87, 93].

One must emphasize that while the nanomaterials fixed on a substrate or those with nanopores do not pose a health risk, the free nanoparticles can become airborne and may be very toxic to human health [21].

1.2.2 Nanomaterials Morphology

A classification of nanomaterials according to their morphology divides them in low- and high aspect ratio particles, see Fig. 2. The high aspect ratio nanoparticles can have different shapes, such as nanowires, nanohelices, nanozigzags, nanopillars, nanotubes, or nanobelts. The low aspect ratio nanoparticles can have many shapes as well, such as spherical, helical, pillar-like, pyramidal, cubes, among others. Most of the images in this chapter illustrate their morphology.

1.2.3 Nanomaterials Composition

According to their composition, nanoparticles can be made of a single material, compact or hollow. Nanomaterials can also be comprised of two or more materials that can be as coatings, encapsulated, barcode, or mixed. More details are given in Sect. 1.1.3a.

1.2.4 Nanomaterials Uniformity and Agglomeration State

According to their uniformity, nanoparticles can be classified as isometric and inhomogeneous. From the point of view of their agglomeration status, nanoparticles can be dispersed or agglomerate. Their agglomeration state depends on their electromagnetic properties, such as surface charge and magnetism. When in a liquid, their agglomeration depends on their surface morphology and functionalization which can confer it either hydrophobicity or hydrophilicity.

1.3 Physicochemical Properties of Nanoparticles

The most important physicochemical aspects of nanoparticles are:

- Composition and surface composition
- Crystalline phase
- Particle size distribution
- Agglomeration/aggregation
- Shape
- Specific surface area
- Roughness/Porosity
- Water solubility/dispersability or hydrophobicity/hydrophilicity
- Zeta potential or surface charge
- Surface chemistry
- Catalytic/Photocatalytic activity
- Magnetic properties
- Optical properties

1.3.1 Composition and Surface Composition

Nanoparticles can be made of a single material or composite materials. Examples of single materials nanoparticles are given in Fig. 3. Many nanoparticles will oxidize in the presence of air and form a thin film at their surface.

Nowadays nanoparticles comprising different composite materials are easily fabricated using a physical vapor deposition process at grazing incidence over a nanoseeded pattern, involving the manipulation of substrate motion and temperature, as demonstrated in Fig. 7 [97, 121]. This technique of deposition at a glancing angle incidence can be used to manipulate the morphology of single materials as well [20].

Figure 7 shows Cu–Ti–Al_2O_3–Au nanohooks, Ag–Al_2O_3–Ag–Al_2O_3–Ag nanobar-codes, Ni–Si–TiO_2–Au nanozigzags [97].

Other examples of various composite nanomaterials are illustrated in Figs. 8, 9, and 10.

Figure 8 shows nanocomposites with a spherical morphology. They are called often encapsulated or core-shell nanoparticles. One can notice core-shells composite nanoparticles: (a) Ni particle encapsulated by a graphite shell [11], (b) graphene sheets encapsulated Si nanoparticle [92], (c), (d) polymer coated cobalt nanoparticles [76], (e), (f) rattle-type Au/CdS composite nanoparticles [143].

Nanocomposites can be fabricated in the most unusual shapes, such as core-shell nanocubes, as shown in Fig. 10. Image (a) shows Pd nanocubes while (b), and (c) shows Pd–Cu core-shell cubes with different shell thickness. [63].

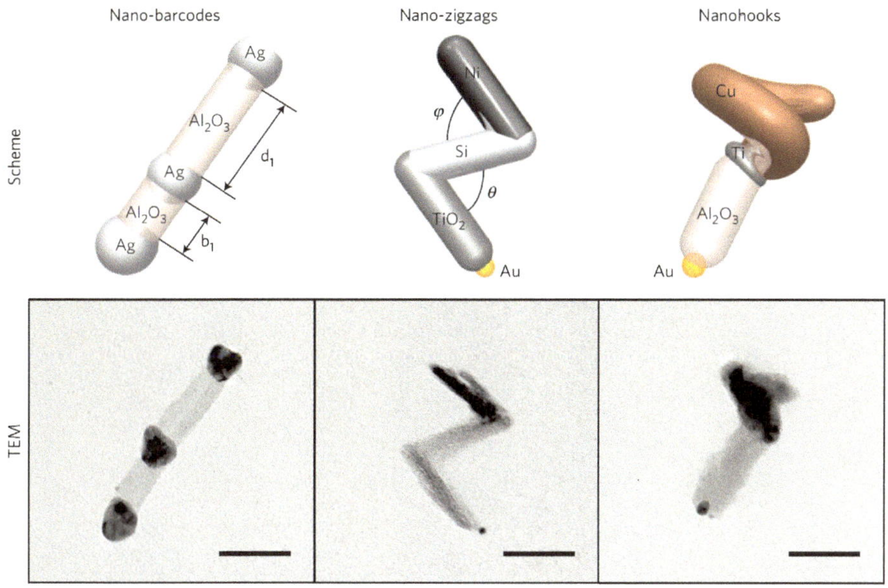

Fig. 7 Composite or hybrid nanoparticles with progressively lower symmetry. Columns from left to right show: nano-barcodes, nanozigzags combining magnetic, semiconducting and insulating materials, and nanohooks with defined chirality. First row, structure models; second row TEM images. Reprinted by permission from Macmillan Publishers Ltd: Nature Materials [97], copyright (2013)

Nanocomposites made of long aspect ratio materials together with spherical small nanoparticles are shown in Fig. 9 where one can notice scanning and transmission electron images of Ag–TiO_2 core-shell nanocomposite nanowires [25].

1.3.2 Crystalline Structure

Crystalline structure is of paramount importance in deciding a nanoparticle's mechanical, chemical, and physical properties. The same material in a different crystalline form can have very different properties and applications.

Examples of nanomaterials of different allotropes are shown in Figs. 11 and 12. Figure 11 shows schematics of the crystalline structure of carbon allotropes: fullerene C60, C70, C540, graphite, single-walled carbon nanotube (SWCNT), amorphous carbon, and multi-walled carbon nanotubes (MWCNT) [15]. Figure 12 shows scanning electron microscopy images of multi-walled carbon nanotubes, graphene, graphite nanoplatelets, and carbon black [93].

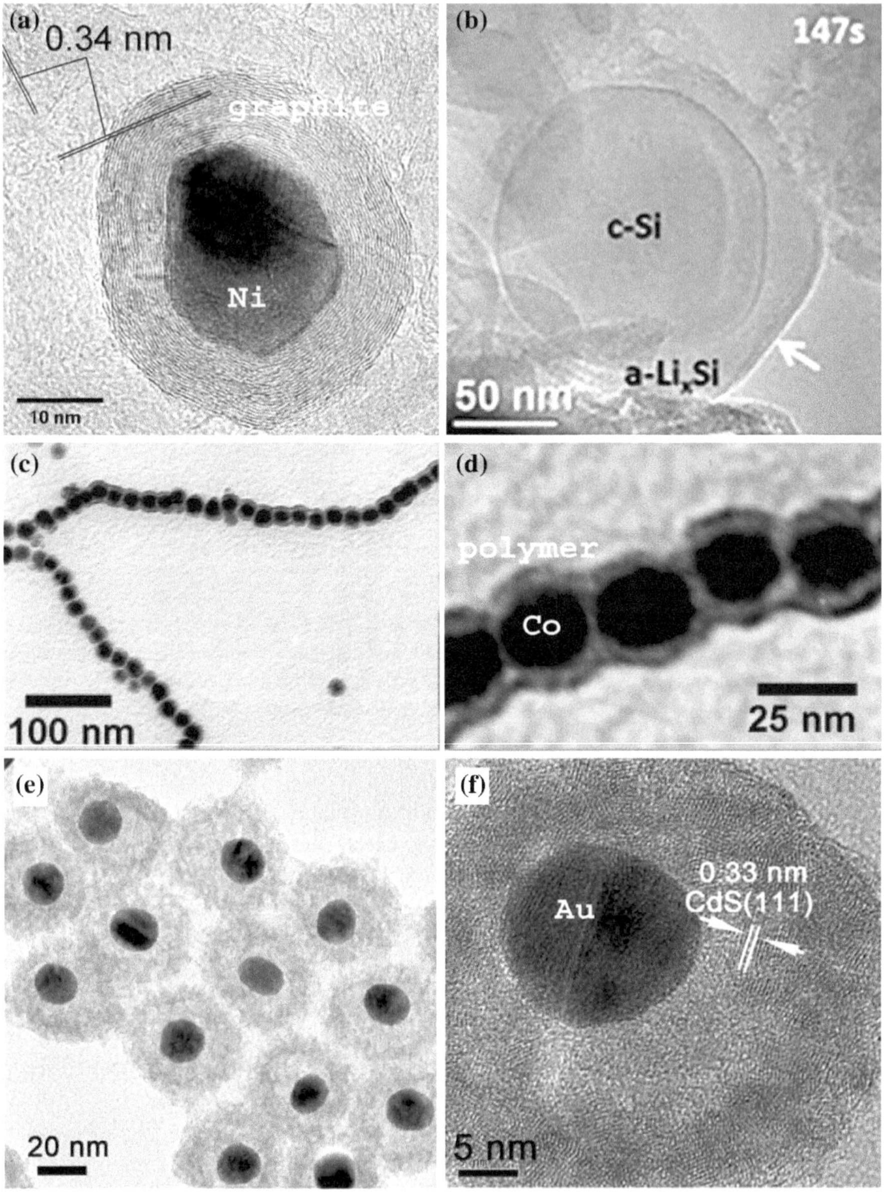
(a)
0.34 nm
graphite
Ni
10 nm
(b)
147s
c-Si
a-Li$_x$Si
50 nm
(c)
100 nm
(d)
polymer
Co
25 nm
(e)
20 nm
(f)
Au
0.33 nm
CdS(111)
5 nm

◀ **Fig. 8** **a** HRTEM image of a Ni particle encapsulated by a graphite shell. The outermost graphene layer represents the shape of the original Ni particle; **b** lithiation of a graphene sheets encapsulated Si nanoparticle; **c**, **d** TEM images of polymer coated cobalt nanoparticles chains at different magnifications. **e**, **f** Rattle-type Au/CdS composite nanoparticles. CdS shells nanostructures encapsulate Au nanocores. Image a reprinted from Carbon, Volume 46, issue 4, R. Anton, On the reaction kinetics of Ni with amorphous carbon, pp. 656–662, Copyright (2008), with permission from Elsevier [11]. Image b reprinted by permission from Macmillan Publishers Ltd: Scientific Reports, [92] copyright (2014). Images c and d adapted with permission from Korth et al. [76]. Copyright (2006) American Chemical Society. Images e and f reprinted from Xia and Tang [143] with permission from John Wiley and Sons

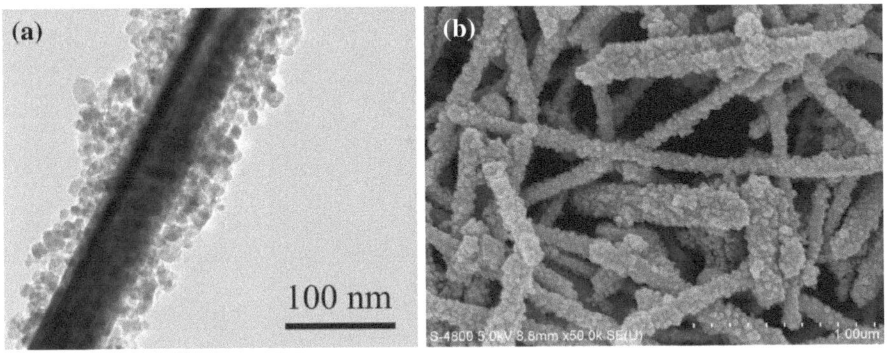

Fig. 9 **a** TEM and **b** SEM image of Ag nanowire coated with TiO_2 nanoparticles. Reprinted from Journal of Hazardous Materials, Volume 177, B. Cheng et al., Preparation and enhanced photocatalytic activity of Ag@TiO2 core–shell nanocomposite nanowires, pp. 971–977, Copyright (2010), with permission from Elsevier [25]

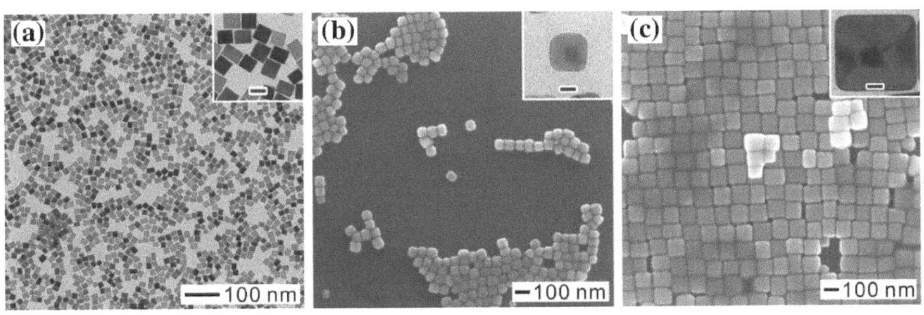

Fig. 10 **a** TEM image of 18-nm Pd nanocubes that served as seeds for the deposition of Cu shells, and **b**, **c** SEM images of Pd@Cu core-shell nanocubes prepared with different volumes of the seed suspension at a concentration of 1.8 mg/mL: 1.5 and 0.5 mL. The edge lengths of the core-shell nanocubes were approximately **b** 50 and **c** 100 nm, respectively. The insets show TEM images of the samples, and the scale bars are 20 nm. Reprinted with permission from Jin et al., ACS Nano, 2012, 6 (3), pp. 2566–2573. Copyright (2012) American Chemical Society [63]

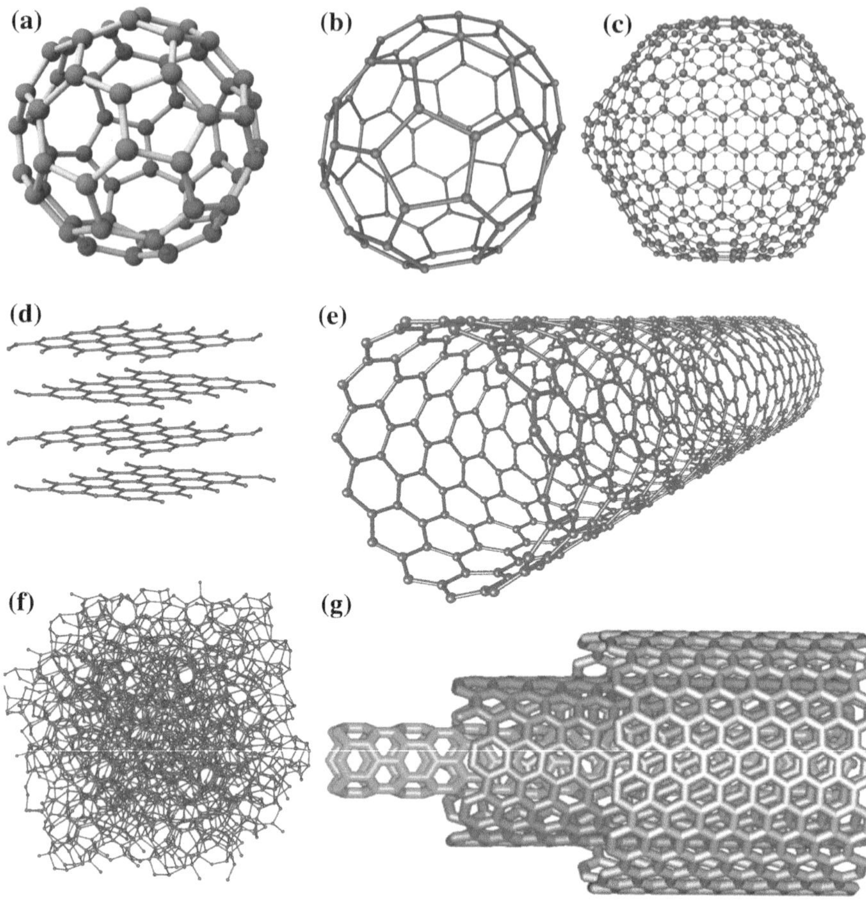

Fig. 11 Structure of various carbon allotropes in nanoform **a** fullerene C60, **b** C70, **c** C540, **d** graphite, **e** single-walled carbon nanotube SWCNT, **f** amorphous carbon, **g** multi-walled carbon nanotubes. Image a courtesy of James Hedberg, images b, c, d, e, and f courtesy of Michael Ströck, **g** reproduced with permission from Balasubramanian and Burghard [15], John Wiley and Sons

1.3.3 Shape or Morphology

Nanoparticles can now be fabricated in almost any shape and containing various combinations of composites materials. Some examples are seen in Fig. 13. Mark et al. demonstrate nanoparticles with complex three-dimensional morphologies comprising a variety of different functional materials [97]. Their fabrication method combines a physical vapor deposition process at grazing incidence over a nano-seeded pattern, involving the manipulation of substrate motion and temperature.

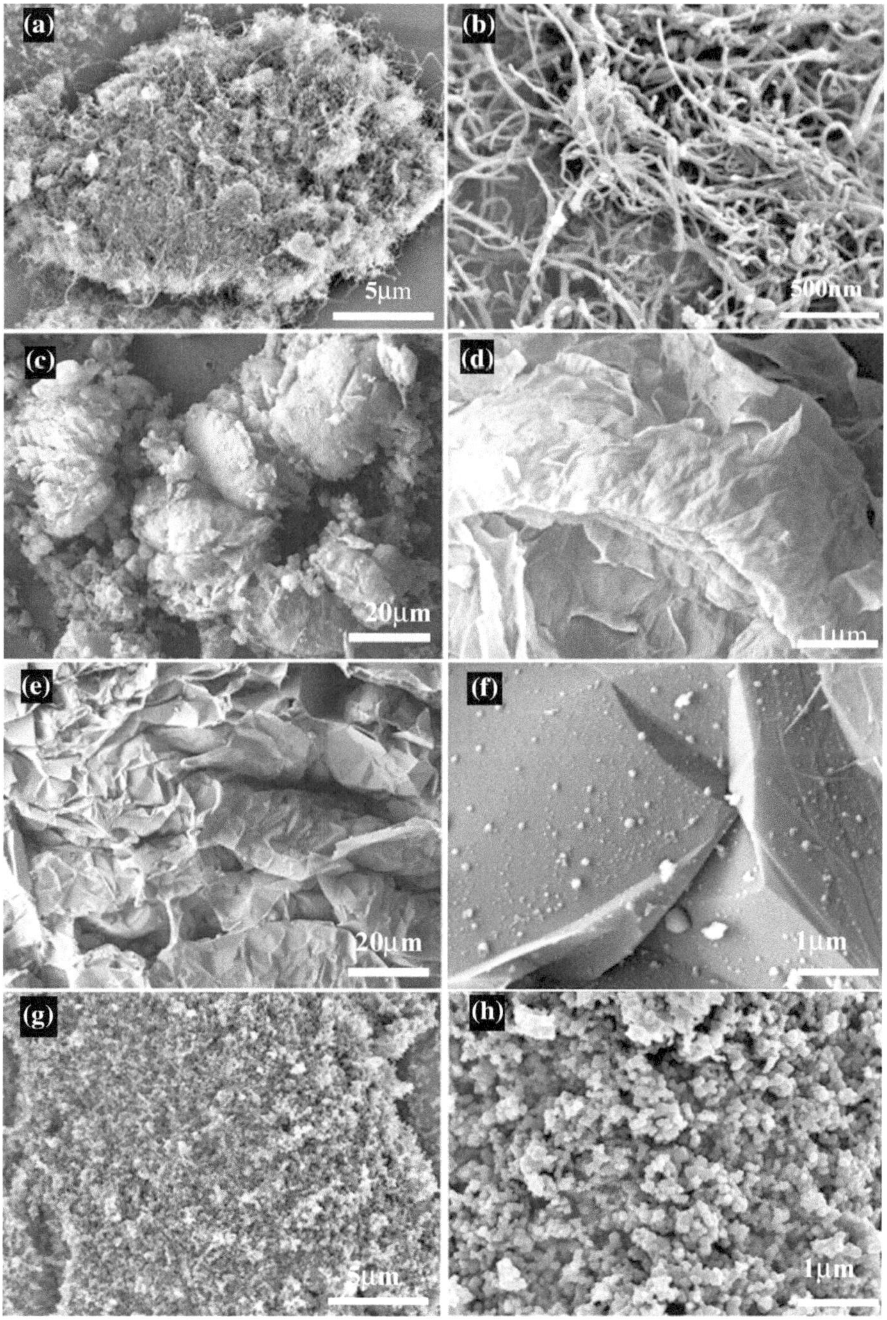
(a)
5μm
(b)
(c)
20μm
(d)
(e)
20μm
(f)
1μm
(g)
(h)
1μm

◀ **Fig. 12** SEM images of carbon allotropes. In the *left side* one can notice the agglomerates while in the *right hand* mode detailed nanostructures are noted at higher magnification; **a**, **b** multi-walled carbon nanotubes, **c**, **d** graphene, **e**, **f** graphite nanoplatelets, **g**, **h** carbon black. Reproduced from Ma-Hock et al. [93]

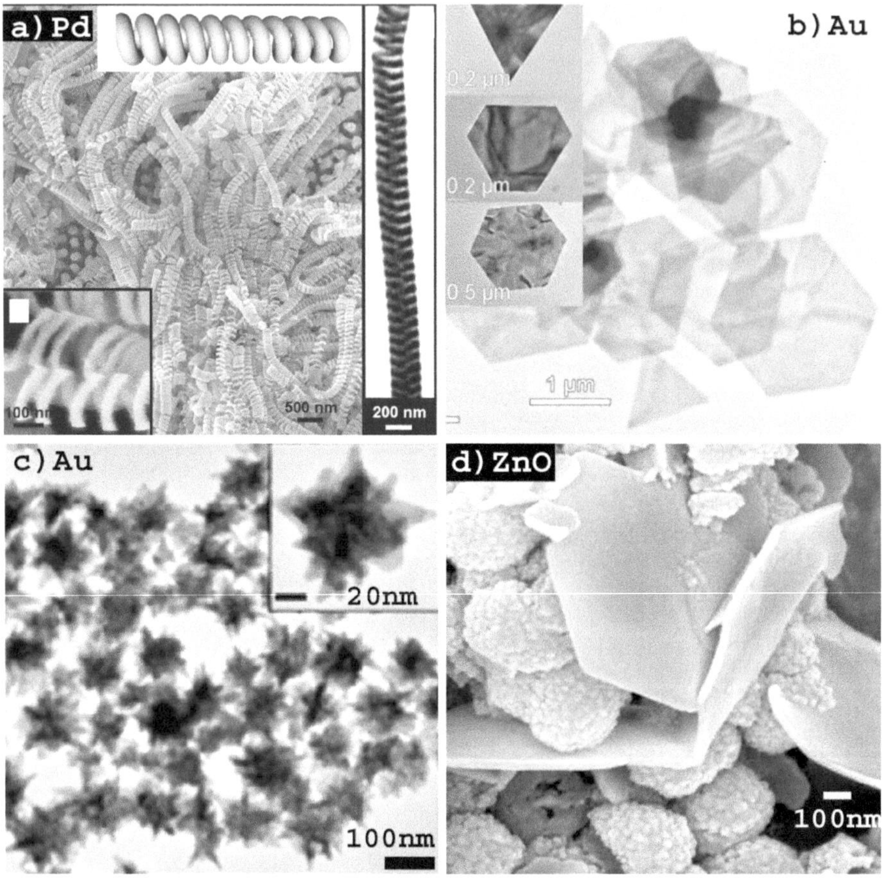

Fig. 13 **a** SEM, TEM images and drawing of Pd nanosprings; **b** TEM micrograph of gold nanoplates. The *inset* shows triangular, truncated triangular, and hexagonal gold nanoplates; **c** SEM image if gold nanostars; **d** field emission scanning electron microscope image of zinc oxide raspberry- and plate-like nanostructures. Image a reprinted with permission from nanoletters, Volume 11, issue 9, Liu et al., Wet-chemical synthesis of Palladium nanosprings, pp. 3979–3982, Copyright (2011) American Chemical Society [87]. Image b reprinted from Xie et al. [144] with permission from John Wiley and Sons. Image a reprinted with permission from J. Phys. Chem. C, 116 (2), Kedia and Kumar, pp. 1679–1686. Copyright (2012) American Chemical Society [65]. Image d reprinted from Journal of Inorganic Biochemistry, Volume 103, Issue 8, M. Z. Hussein, Bacillus cereus as a biotemplating agent for the synthesis of zinc oxide with raspberry- and plate-like structures, Copyright (2009), with permission from Elsevier [60]

Nanoparticles with different morphologies have been demonstrated, such as:

- Nanorods [120, 19, 97],
- nanozigzags [120, 97],
- nanohooks [97],
- nanostars [66, 65],
- nanocubes [63],
- nanohelices [120, 97, 65, 87, 139] and
- nanoplates [144, 60]

Various materials have been deposited in nanohelical: such as Pd [87], Au, Ag [97], Ni/Ti [139]. To have an idea of the size and morphology of nanohelixes, in reference [97] is described the fabrication of gold nanohelixes that have two turns, a pitch of 34 nm and an overall height of 100 ± 8 nm.

Long aspect ratio nanoparticles encompass not only carbon nanotubes, but Si nanowires [83], Ag nanowires [25], Pd nanotubes [87]. Some of them have already been discussed, as seen in Fig. 6.

More complex morphologies are also attainable by different synthesis methods, such as core-shell nanoparticles: spherical CdSe/CdS core-shell nanoparticles [143], Ni encapsulated by a graphite shell [11], Pd–Cu core-shell cubes [63], hollow $BaTiO_3$ nanoparticles [85]. An example is given in Fig. 14 showing $BaTiO_3$ supercages [85].

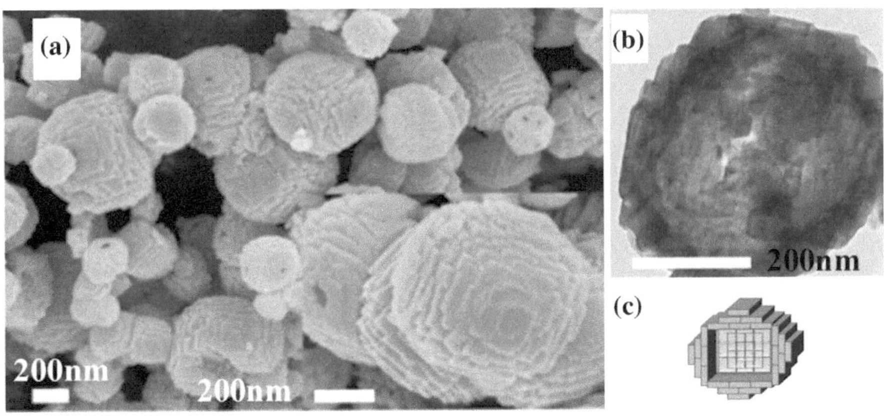

Fig. 14 $BaTiO_3$ supercages **a** SEM images, **b** TEM images, **d** illustration. *Scale bars* 200 nm. Adapted with permission from ACS Nano, vol. 9, issue 1, $BaTiO_3$ Supercages: Unusual Oriented Nanoparticle Aggregation and Continuous Ordering Transition in Morphology, Li, pp. 496–502. Copyright (2015) American Chemical Society [85]

1.3.4 Surface Area

Particle size and surface area are important parameters that play a major role in interaction of nanomaterials with the exterior. Decreasing the size of the nanoparticle leads to an exponential increase in surface area relative to volume, resulting in an enhanced reactivity. See Fig. 1 that shows the variation of number of surface and bulk atoms with the nanoparticle size.

1.3.5 Water Solubility-Dispersability or Hydrophobicity-Hydrophilicity

Nanoparticles can be insoluble or repelling water, also called hydrophobic. Water-soluble nanoparticles are called hydrophilic.

Many of the engineered nanoparticles are hydrophobic and do not form stable suspensions in aqueous media. They are nanoparticles made of metals and their compounds (such as titanium dioxide or quantum dots), nonmetals (carbon-based nanoparticles including fullerenes C60, carbon nanotubes, carbon soot, and silicon-based nanoparticles).

In order to improve their dispersability via electrostatic repulsion, nanoparticles can be coated with different substances. For example, gold nanoparticles can be functionalized with citrate groups, while multi-walled carbon nanotubes are usually functionalized via a selective breaking of C = C bonds achieved through oxidation [112].

The soluble or biodegradable nanoparticles may convert into molecular components as a result of biological and/or photochemical decomposition.

1.3.6 Catalytic/Photocatalytic Activity

Nanoparticles that have surface facets sufficiently large will have catalytic properties that approximate those of their bulk counterparts with crystalline surfaces [9]. With decreasing size, usually smaller than 3 nm, a nanoparticle catalytic activity starts to deviate from its bulk behavior due to the increasing contribution of the corner and edge atoms of the nanoparticle. Spherical nanoparticles with a size of about 3 nm have half of their atoms on the surface. In the case of nanoparticles with sizes between $\sim$1 and 2 nm, even slights changes in its structure, the deletion or addition of a few atoms, can have an important impact on its catalytic properties. Nanoparticles with such small size are usually hard to characterize experimentally due to the fact that their structure may be altered during catalytic reactions [9].

Furthermore, for a given crystalline structure, the different crystalline surface may show a different reactivity toward the same reaction. It has been shown that different crystal surfaces of Pt have very different electronic structures and atomic arrangements, consequently having a different reactivity toward the same reaction [111].

1.3.7 Magnetic Properties of Nanoparticles

There are two types of materials that exhibit magnetism in nanoform. There are magnetic nanomaterials that are magnetic in bulk form and nanoparticles that show magnetic behavior only in nanoform, while in bulk are nonmagnetic. Below we will discuss the two classes of magnetic nanomaterials.

Magnetic properties of nanoparticles that are magnetic in bulk form

Magnetic materials are defined as materials that change their magnetic behavior as a result of an external magnetic field. There are five main types of magnetic materials: ferromagnetic, paramagnetic, diamagnetic, antiferromagnetic, and ferromagnetic, as shown in Fig. 15a [61].

In ferromagnetic materials (such as Fe, Ni, Co) the magnetism arises due to atomic net magnetic moments as a result of unpaired electrons [59]. The material has domains containing large numbers of atoms with parallel magnetic moments, each domain having a net magnetic moment pointing in a specific direction [61]. Due to the random orientation of the magnetic moments of the domains, the net magnetic moment of the material is zero. In an external magnetic field, the domains of the ferromagnetic material align along the direction of the applied magnetic field, resulting in a large net magnetic moment. After the removal of the external field there is a residual magnetic moment left.

In paramagnetic materials (Gd, Mg, Li, Ta) magnetic domains are absent even though atoms have net magnetic moment due to unpaired electrons [61]. After the application of an external magnetic field, the atoms align along the direction of the field resulting in a weak net magnetic moment. After the removal of the field, the paramagnetic materials do not have a remanent magnetization.

Diamagnetic materials (bulk Cu, Ag, Au, and most of the rest of the elements) have atoms with no unpaired electrons and show a zero net magnetic moment [61]. Their magnetic response in an external magnetic field is very weak and after the removal of the field they do not retain any magnetic moment.

Antiferromagnetic materials (MnO, CoO, NiO) have two different atom types occupying different lattice positions [61]. The two types of atoms have opposite magnetic moments, resulting in zero net magnetic moment.

Ferrimagnetic materials (magnetite-Fe_3O_4) have also different atoms types occupying different lattice sites with antiparallel magnetic moments, but different magnitudes. Hence the magnetic moments do not cancel out, which results in a net

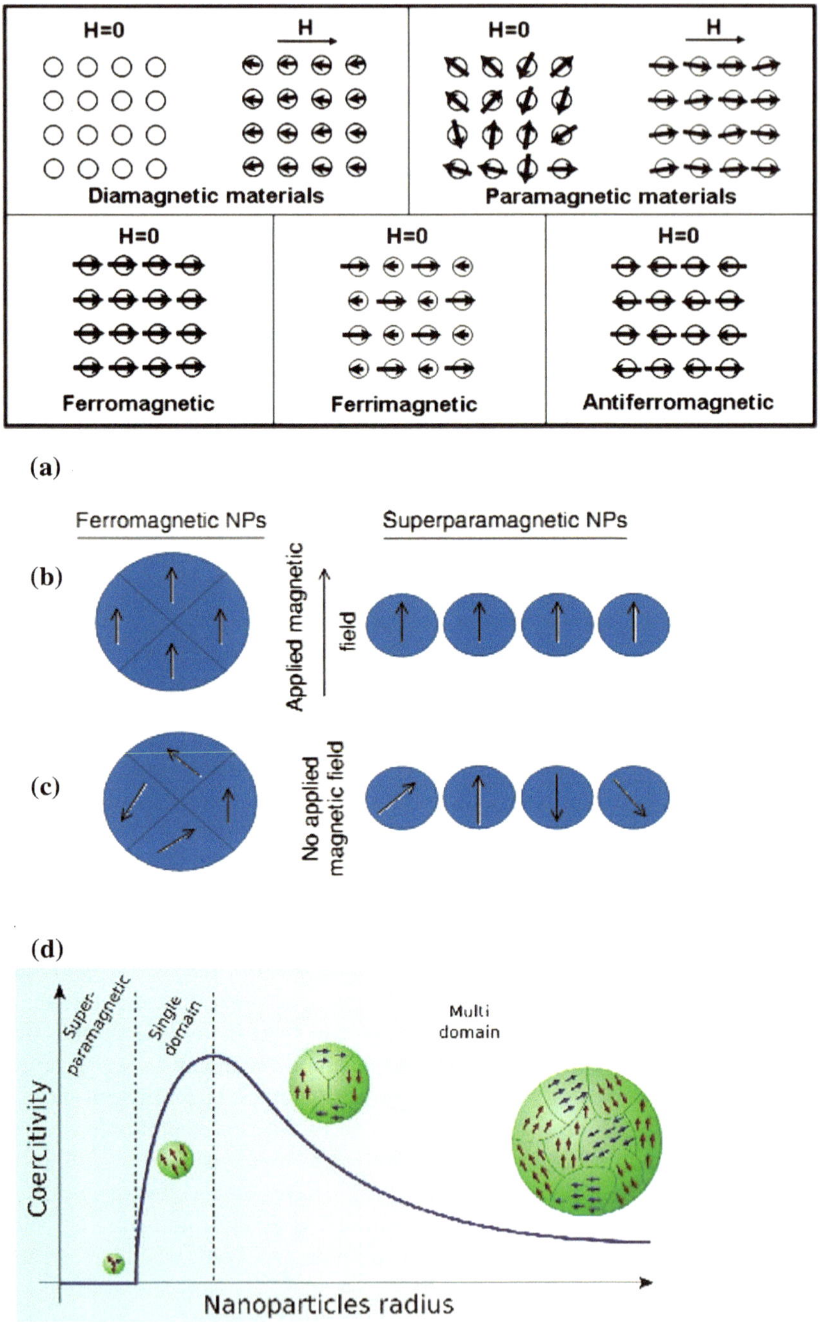
H=0
H
Diamagnetic materials
H=0
H
Paramagnetic materials
H=0
Ferromagnetic
H=0
Ferrimagnetic
H=0
Antiferromagnetic
(a)
Ferromagnetic NPs
Superparamagnetic NPs
(b)
Applied magnetic field
(c)
No applied magnetic field
(d)
Super-paramagnetic
Single domain
Multi domain
Coercitivity
Nanoparticles radius

◀ **Fig. 15** **a** The behavior of magnetic dipoles in the absence and presence of external magnetic fields. According to their behavior they can be classified as diamagnetic, paramagnetic, ferromagnetic, ferrimagnetic, antiferromagnetic [62]. **b**, **c**, **d** The magnetization of ferromagnetic and superparamagnetic nanoparticles under an external magnetic field; **a** under an external magnetic field the domains of a ferromagnetic nanoparticle align with the applied field and the magnetic moment of single domain superparamagnetic nanoparticle aligns with the applied field; **b** In the absence of an external field the ferromagnetic nanoparticles will have a net magnetization and superparamagnetic nanoparticles will have no net magnetization due to rapid reversal of the magnetic moment; **c** schematic illustration of the coercivity-size relations of small particles. Reprinted from Akbarzadeh et al. [5]

spontaneous magnetic moment. In an external magnetic field, both antiferromagnetic and ferrimagnetic materials behave similarly to ferromagnetic materials.

Magnetic nanoparticles are usually classified depending on their magnetic response in an external magnetic field (Fig. 15a). According to this, nanoparticles can be either ferrimagnetic, ferromagnetic, or superparamagnetic particles (a special case of ferro- or ferrimagnetic particles) [74, 62]. Examples of magnetic nanoparticles are Co, Ni, Fe_3O_4, $CoFe_2O_4$, $HoMnO_3$, $MnFe_2O_4$, FePt [74].

Below a specific critical diameter, magnetic nanoparticles show magnetic responses similar to paramagnetic materials, i.e., in the absence of an external magnetic field show a zero average magnetic moment, while in an external field increasing magnetic moments [74]. This phenomenon occurs due to comparable or larger thermal fluctuations than the energy barrier for moment reversal. When the magnetization of nanoparticles in the absence of an external field is zero, then these nanoparticles are called superparamagnetic. Superparamagnetic nanoparticles are used in applications such as drug delivery or magnetic resonance imaging.

Figure 15b, c shows the magnetization of ferromagnetic and superparamagnetic nanoparticles under an external magnetic field. Under an external magnetic field the domains of a ferromagnetic nanoparticle align with the applied field and the magnetic moment of single domain superparamagnetic nanoparticle aligns with the applied field. In the absence of an external field, the ferromagnetic nanoparticles will have a net magnetization and superparamagnetic nanoparticles will have no net magnetization due to rapid reversal of the magnetic moment [5].

Magnetic properties of nanoparticles that are not magnetic in bulk form

In addition to the above-discussed magnetic nanoparticles, some nanoparticles of materials which are not magnetic in the bulk form exhibit a magnetic behavior in nanoform, for nanoparticle size smaller than a few nanometers:

- Au [54, 55, 145, 94, 84, 104],
- Pt [124],
- Pd [54, 130],
- HfO_2 [146].

Magnetism has also been observed not only in Au nanoparticles, but in thin films of Au with 27 nm thickness deposited on Pyrex glass [119].

Magnetic behavior can be tuned by surface coatings or ligands [80, 31]. The magnetic anisotropy of Pt nanoparticles is tunable via the modification of surface conditions (thiol coatings) [124].

While at some point it was believed that the surface coatings or impurities were responsible for the observed magnetism of these noble metals nanoparticles [145], subsequently it was shown that also noble metal nanoparticles prepared under high vacuum conditions and with high purity show magnetic behavior [126, 130].

In addition, it has been shown that ligand-stabilized Au nanoparticles exhibit different magnetic behaviors: diamagnetic, paramagnetic, or ferromagnetic, as shown in Fig. 16 [80]. Clusters with the same core size of about 25 atoms but different ligands show two different magnetic behaviors. Mixed ligand-stabilized and thiol-stabilized clusters of 25 and 38 Au atoms are diamagnetic while clusters with 55 atoms are ferromagnetic.

While it could be argued that the influence of a substrate might influence the magnetism of nanoparticles, it was shown that freestanding Pd nanoparticles with clean surface exhibit ferromagnetism [130]. The ferromagnetic behavior is due to the ferromagnetic ordering in a few topmost surface layers, while the rest of the nanoparticle stays paramagnetic, similar to Pd bulk [130]. Consequently, the origin of the ferromagnetism Pd small clusters can be associated with the two-dimensional character of the particle surface.

Similarly, it has been observed that the surface atoms of Au nanoparticles are ferromagnetic while the core is paramagnetic [145].

While in the bulk form Au is chemically stable and with diamagnetic properties, at the nanoscale below a threshold size, gold nanoparticles exhibit ferromagnetism [55]. The curve of the magnetization versus nanoparticle diameter exhibits a peak for a nanoparticle size of around 3 nm [55]. The magnetism reverts back to the one of bulk for nanoparticles larger than 4 nm. The magnetization peak can be

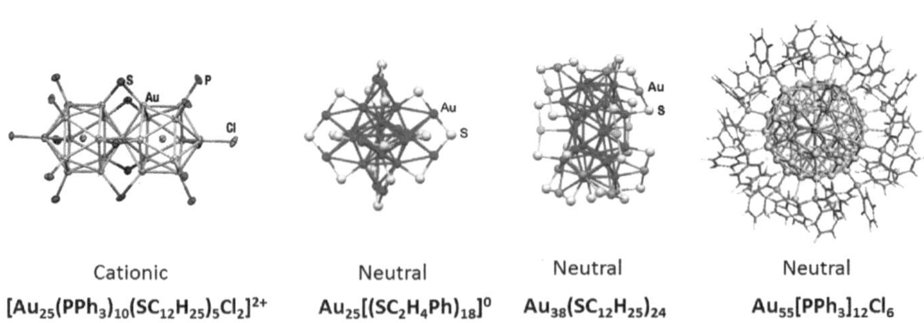

Fig. 16 Ligand-stabilized atomically precise gold clusters (APGCs) with their experimentally observed magnetic behavior. Reproduced with permission from Krishna et al. [80], John Wiley and Sons

explained by the decrease in surface magnetism and increase in spin-polarized electron number with the nanoparticle diameter.

Some authors explain the observed magnetism of Au nanoparticles (and possibly other metals) in the existence of self-sustained persistent currents having an orbital origin [46].

The combination of quantum-size and Coulomb-charging effects cause the properties of small metallic particles to depend on whether they have an odd or even number of electrons [141]. However, some authors suggest that this mechanism cannot explain the observed magnetization of Pd and Au nanoparticles with a diameter of 3 nm, which shows an unexpectedly large magnetic moment of about 20 spins per particle [54]. They suggest that their results might be explained by the existence of some common and characteristic spin correlation mechanism in the nanoparticles [54]. At a critical size of about 3 nm diameter, there is a transition from an atomic spin correlation region to a metallic spin correlation region.

1.3.8 Optical Properties

The optical properties of nanoparticles vary with their composition, morphology, and size [69, 97, 99, 128].

Optical properties of nanoparticles pertain to the way that they interact with light. If one exposes matter to light, several processes can occur [58]:

- Absorption of light
- Scattering of light at the same frequency as the incoming light (Mie or Rayleigh scattering)
- Reemission of the absorbed light (fluorescence)
- Enhancing of the local electromagnetic field of the incoming light, thus enhancing spectroscopic signals from molecules at the material surface.

Au nanoparticles exhibit an enhancement of all the above-mentioned processes due to a unique interaction of the nanoparticle-free electrons and light [58]. When a gold nanoparticle is exposed to an external optical field, the electrical component of the light acts upon the conduction-band electrons and displaces them, creating uncompensated charges at the nanoparticle surface [58, 69]. Because the polarization occurs at the nanoparticle surface, these oscillations are called surface plasmons. Plasmons are defined as collective excitations of conductive electrons in metals [69]. There are several types of plasmons: bulk plasmons (or 3D plasma), surface propagating plasmons polaritons (in 2D films), and surface localized plasmons (in nanoparticles).

The surface plasmons have a well-defined resonance frequency, the coherent oscillations of the nanoparticle electrons in resonance with the electromagnetic field is called surface plasmon resonance (SPR).

In the case of Au nanospheres, the surface plasmon resonance is in the visible spectrum at around 520 nm, hence the red color of solutions of spherical Au nanoparticles, as shown in Fig. 17 [58, 117]. The resonance of gold nanorods is

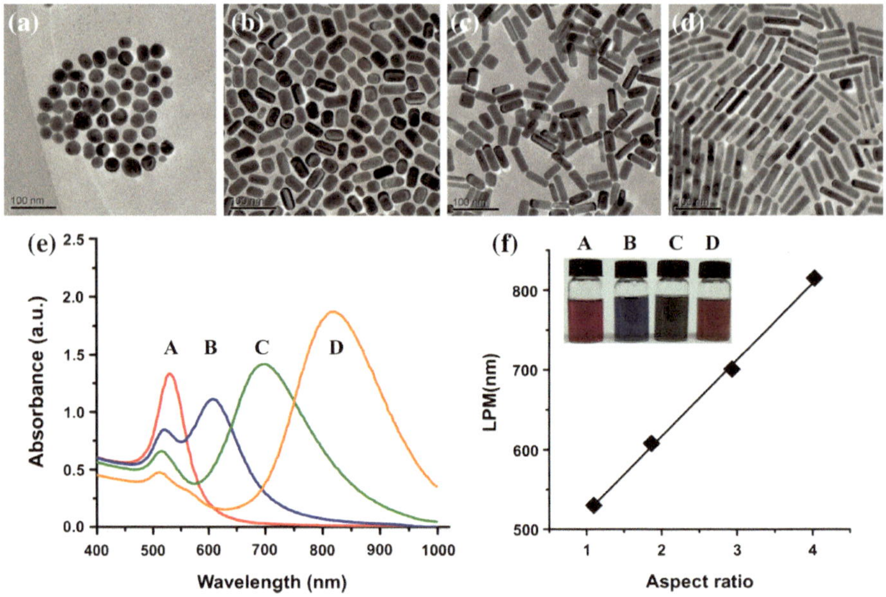

Fig. 17 TEM images of cetyltrimethylammonium bromide-coated Au nanoparticles with different aspect ratio (E) representative UV-Vis-NIR absorption spectra. (F) The relationship between longitudinal plasmonic maximum LPM and calculated aspect ratios. The inset is the photograph of the Au NRs suspensions. Adapted from Qiu et al. [117]. Adapted from Biomaterials, Volume 31, issue 30, Qiu et al. Surface chemistry and aspect ratio mediated cellular uptake of Au nanorods, pp. 7606–7619, Copyright (2010), with permission from Elsevier

created by oscillations along both the long and short axis, resulting in a stronger band in the near-infrared region and a smaller band in the visible region, respectively.

These properties make Au attracting in biomedical applications [58]. For example, biological tissue labeled with gold nanoparticles can be visualized clearly by the surface plasmon resonance scattering of the Au nanoparticles under monochromatic light from a laser. The light scattering from Au nanoparticles can be 10^5 stronger than the light scattering from a fluorescein molecule, widely used as a fluorescent tracer [58]. For highly scattering tissue, the fluorescence of Au nanorods can be used for imaging. The fluorescence of Au nanorods is about 6–7 times higher than that of bulk Au, and it is attributed to the excitation of the longitudinal plasmon [58]. The luminescence of Au nanorods with a length of more than 200 nm can be seen with the naked eye.

Quantum dots are also used for their photoluminescence properties. The photoluminescence of a quantum dots can be manipulated to specific wavelengths by controlling particle diameter as shown in Fig. 18a, b [133]. Examples of images of quantum dots are shown in Fig. 18d, e, and f [148, 149].

Quantum dots are semiconducting nanocrystals of approximately spherical shape with a typical diameter between 2 and 8 nm, encompassing between 200 and 10,000 atoms [7]. In a bulk semiconductor the valence electrons can be excited to a

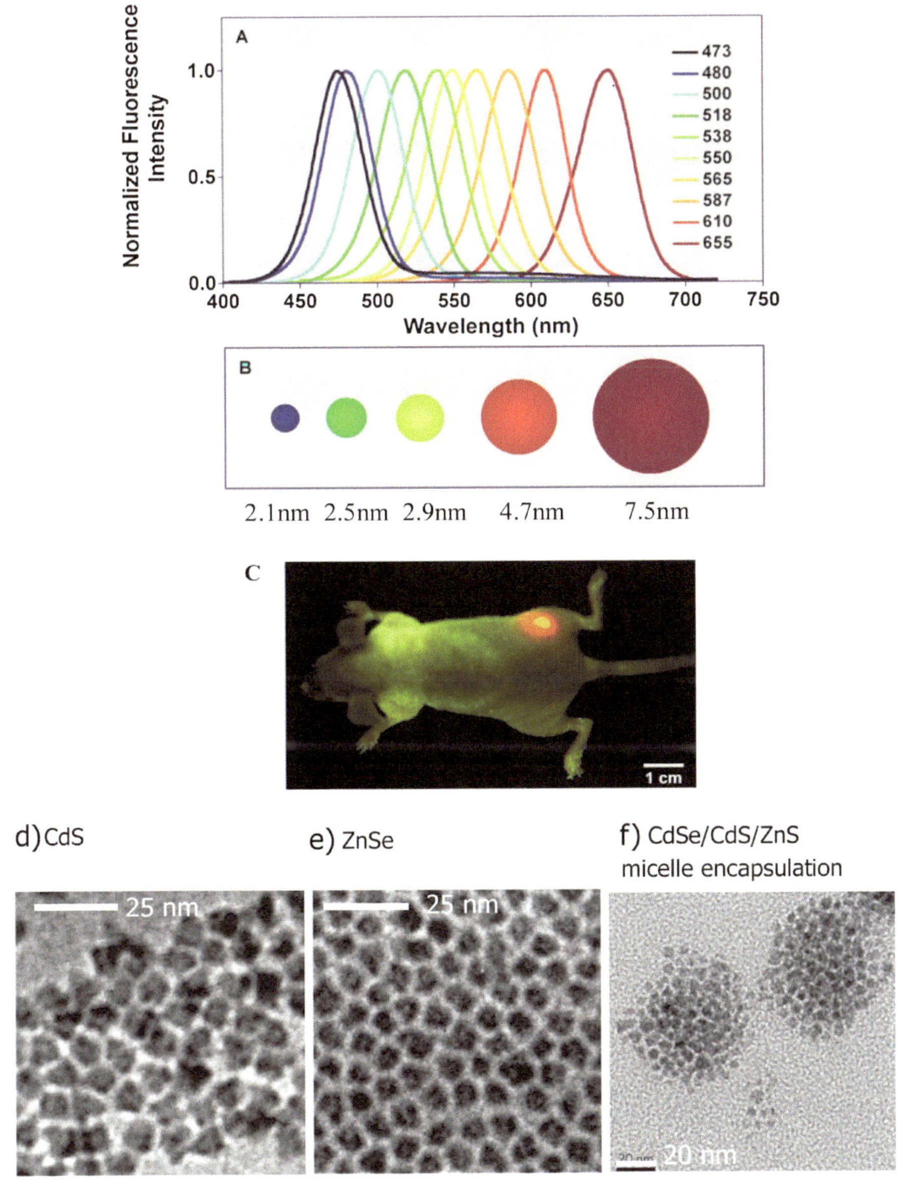
A
Normalized Fluorescence Intensity
1.0
0.5
0.0
400
450
500
550
600
650
700
750
Wavelength (nm)
473
480
500
518
538
550
565
587
610
655
B
2.1nm 2.5nm 2.9nm 4.7nm 7.5nm
C
1 cm
d) CdS
25 nm
e) ZnSe
25 nm
f) CdSe/CdS/ZnS micelle encapsulation
20 nm

◀ **Fig. 18** (Color online) **a** The fluorescence spectra variation with the size of CdSe quantum dots, **b** illustration of the relative particle sizes, **c** in vivo tumor imaging with quantum dots. Fluorescence image of a live mouse with QD-tagged cancer cells injected into its right flank, showing an orange glow (Courtesy of Dr Xiaohu Gao, Emory University). **d**, **e**, **f** Transmission electron microscope images of quantum dots. **d** CdS-rich quantum dots (scale bar, 25 nm), **e** ZnSe-rich quantum dots. **f** CdSe/CdS/ZnS quantum dots after phospholipid micelle-encapsulation. Images a, b, c are adapted from Smith and Nie [133] with permission of The Royal Society of Chemistry. Images d and e adapted by permission from Macmillan Publishers Ltd: Nature Photonics [148], copyright (2015). Image f adapted by permission from Macmillan Publishers Ltd: Nature Nanotechnology, [149], copyright (2012)

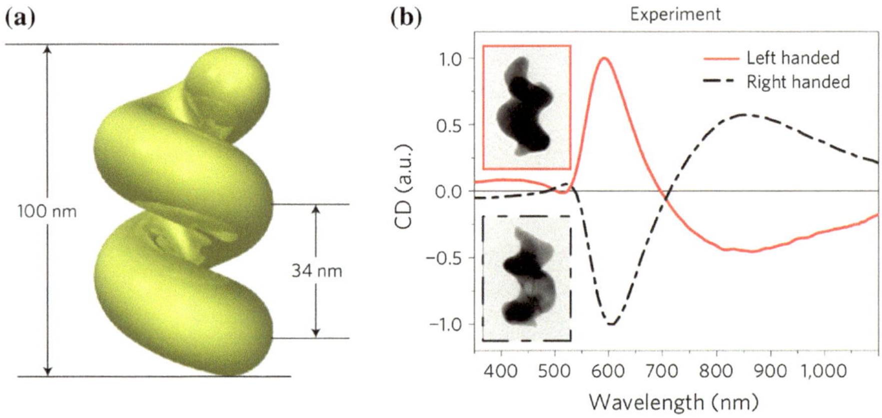

Fig. 19 (Color online) **a** Schematics of Au nanohelix with two turns showing critical dimensions, **b** chiro-optical response of solutions of Au nanohelices; normalized circular dichroism (CD) spectra of left-handed and right-handed helices. *Inset* TEM images of grown structures with *left* (*top*) and *right* (*bottom*) chirality (image dimensions: 85 nm × 120 nm). Reprinted by permission from Macmillan Publishers Ltd: Nature Materials [97], copyright (2013)

higher energy conduction band, the difference in this energy being called the band gap energy of the semiconductor [133]. After the emission of a photon with energy equal to the bang gap energy, an excited electron may relax to its ground state. The band gap of a semiconducting nanoparticle is dependent on nanoparticle size, increasing with the nanoparticle size [133]. As a result the emitted light has shorter wavelengths. Biological applications of quantum dots are dictated by the fact that their size-tunable properties occur at the same size regime as biological molecules such as proteins and nucleic acids.

Optical properties of nanoparticles can also be modified as a result of their chirality or handnesness [97]. Figure 19 illustrates the tuning of the chiro-optical response with the change in nanoparticles morphology. The long-wavelength circular dichroism spectra of the nanohelices with right-handed chirality show a positive ellipticity, while for left-handed ones they exhibit a negative one. The spectral peak position is a function of the radius of the helix.

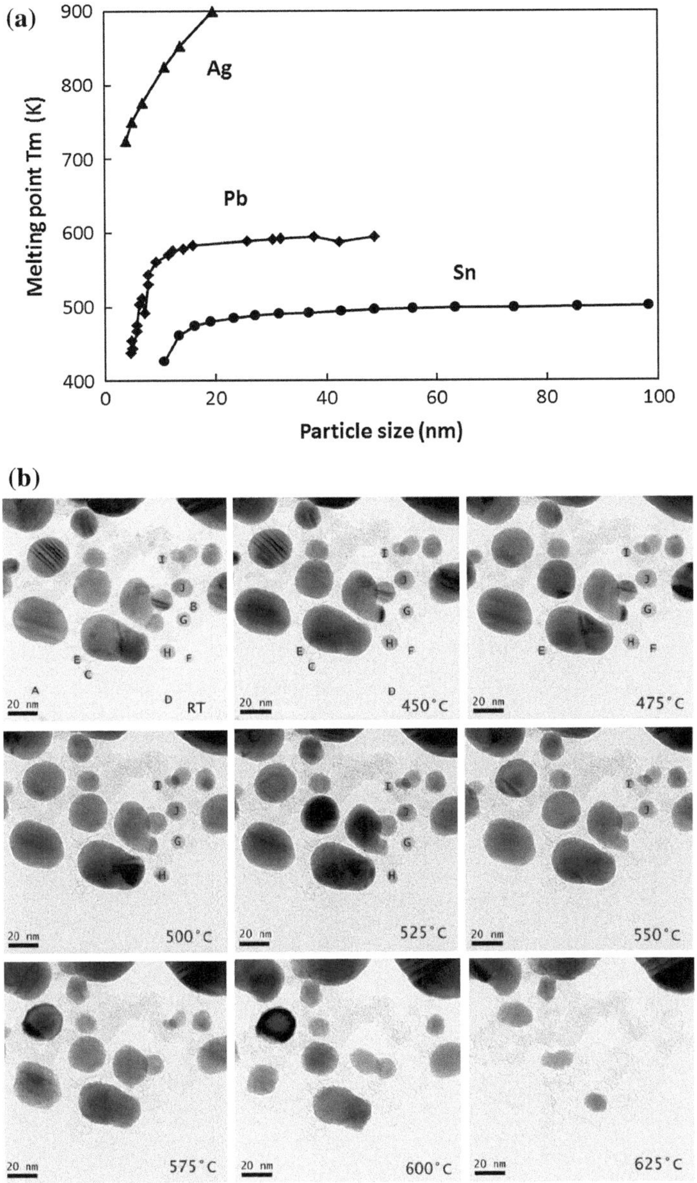

Fig. 20 **a** The melting point, T_m, versus particle size for Ag [13], Sn [82], and Pb [116]; **b** sequence of TEM images exhibiting melting and vaporization of Ag nanoparticles. Image b reprinted with permission from Cambridge University Press [13]

1.3.9 Mechanical Properties—Melting Temperature

Melting temperature of nanoparticles

Large cluster nanoparticles

The melting temperature decrease with decreasing particle size is a thermodynamic consequence of the change in the surface-to-volume ratio that is confirmed experimentally [3, 13, 30, 82, 116, 150]. For example, the melting temperature of gold nanoparticles with a diameter of 3.8 nm (approximately 1,700 atoms) is about 1000 K, while for 2.5 nm nanoparticles (about 500 atoms) is around 500 K, and the bulk melting point is 1,337 K [3].

As illustrated in Fig. 20, the melting point decreases smoothly with particle size [3, 13, 82, 116].

Small clusters nanoparticles

Lowering further the particle size, in the cluster-size regime (tens to several hundreds of atoms), the simple addition or removal of a single atom can drastically change the physical properties of the cluster [115]. While the decrease of melting temperature with size is true for large particles, when lowering the particle size below a certain threshold, the melting temperature starts showing size-dependent fluctuations [3]. This behavior is illustrated in Figs. 21 and 22. The measurements of the melting temperature for Na clusters with n = 55–360 atoms show a fluctuating dependence with cluster size. The melting temperatures are depressed by almost 40 % compared to the Na bulk, and the average trend is for the melting temperature to decrease with increasing cluster size (Adapted from Aguado and

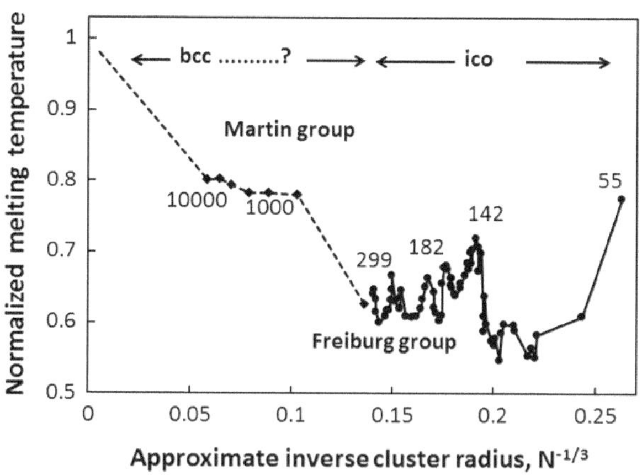

Fig. 21 Normalized melting temperatures measured for Na clusters as a function of the inverse cluster radius, $N^{-1/3}$. The normalized melting temperature is the cluster melting temperature normalized to the bulk melting temperature. The smaller size clusters have an icosahedral packing (ico). At an unknown critical size must be a transition toward bulk-like (bcc) structures. Adapted from Aguado and Jarrold [3] and references therein

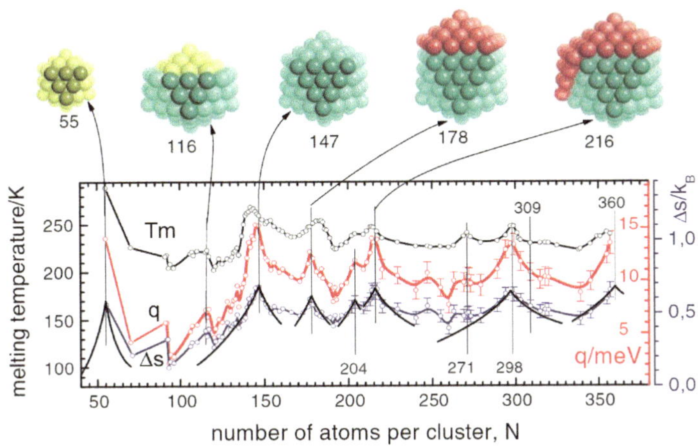

Fig. 22 (color online). Upper panel: Na clusters of icosahedral growth pattern. *Lower panel* Size dependence of the melting temperature *(black)*, the latent heat of fusion per atom (q, *red*), and the entropy change upon melting per atom (Δs, *blue*). The three data have their maxima around the same N. The cluster sizes are indicated for some peaks. Melting temperatures of Na clusters exhibit size-dependent fluctuations that correlate with geometrical shell closings. (Adapted from reference Haberland et al. [49]). Reprinted image with permission from Haberland et al. Physical Review Letters 94, 035701, 2005, Copyright (2005) by the American Physical Society, http://link.aps.org/doi/10.1103/PhysRevLett.94.035701

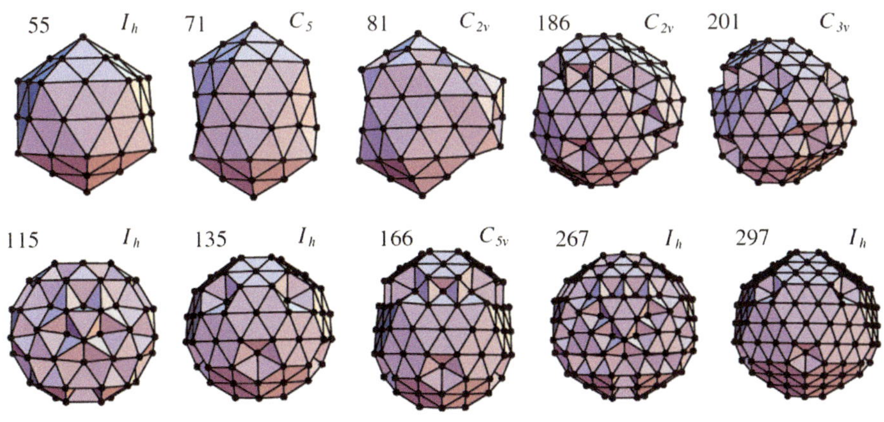

Fig. 23 A selection of Na cluster global-minimum structures that have enhanced stabilities. The structures are labeled by their size and point group symmetry. Adapted from Noya et al. [107]. With kind permission from Springer Science + Business Media: European Physical Journal D, Geometric magic numbers of sodium clusters: Interpretation of the melting behavior, 43, 2007, 57–60, Noya et al. Copyright EDP Sciences, Societa Italiana di Fisica, Springer-Verlag 2007

Jarrold [3]). For clusters with larger size, having between 1,000 and 10,000 atoms the melting temperature is depressed by only 20 % compared to the bulk, and there is no clear extrapolation trend to the bulk value. The small sizes clusters are based on Na icosahedral packing (ico), while the larger ones are Na bulk-like (bcc) structures. The size-dependent fluctuations in the melting temperatures of Na clusters correlate with geometrical shell closings (see Fig. 22) [49]. Clusters that are more compact melt at a higher temperature [3, 4]. Figure 23 shows a selection of Na cluster structures that have enhanced stabilities.

Calorimetric measurements of the heat capacity versus temperature of metallic clusters show a peak in the heat capacity due to latent heat, which indicates the value of the melting temperature of the clusters, as shown in Fig. 24a [115]. The deduced melting temperature for these metals shows size-dependent fluctuations, as illustrated in Fig. 24b [115]. One can notice large fluctuations (several 100°) can occur just by simply adding or removing a single atom. For some clusters, the heat capacities have extra features (an additional peak or a dip) that are related to a more complex behavior than simple melting. For small clusters the influence of the substrate might perturb their properties, and also for unsupported clusters it is difficult to determine when they melt. Three main causes can lead to a second feature in the heat capacity versus cluster size: partial melting when a part of the

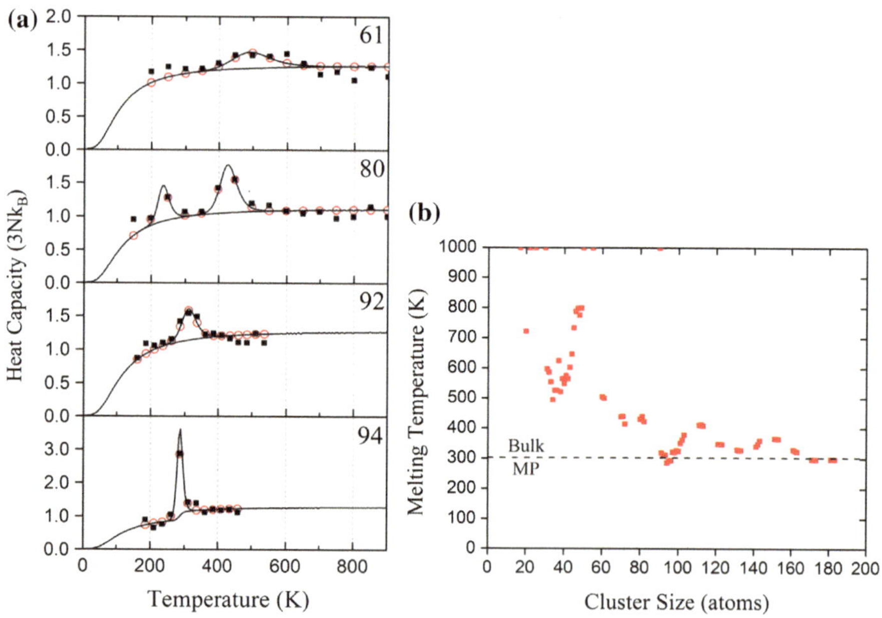

Fig. 24 **a** The heat capacity versus temperature for very small Ga clusters with various number of ions: 61, 80, 92, 94. Here 3 N = 3n − 6 + 3/2, where n is the number of atoms in the cluster, and k_B is the Boltzmann constant. **b** Melting temperatures determined from the fits to the peaks in the heat capacities versus the number of atoms in each cluster. Reprinted with permission from Pyfer et al. [115]. Copyright (2014) American Chemical Society

cluster melts at a lower temperature than the rest, the existence of two structural isomers with different melting temperatures, and a solid-to-solid transition followed by melting [3].

In this size regime, it is possible that the melting temperature of very small clusters to be higher than the melting temperature of bulk, due to the fact that the structure of the clusters is different than the structure of the bulk [131]. It has been shown that Sn nanoparticles (cluster ions) with 10–30 atoms have remained solid at 555 K, which is 50 K above the melting point of bulk Sn [131].

The exact mechanism by which the melting points become elevated is still under research. Some authors calculated by using first principles parallel tempering molecular dynamics in the microcanonical ensemble that a strong charge segregation between the internal and surface atoms might be responsible for a greater than bulk melting temperature for Al clusters composed of 20 atoms [108]. The clusters show negatively charged internal atoms and positive charge at the surface. Other authors suggested that there is a marked shortening and stiffening of atomic bonds with decreasing the number in the atomic clusters [114]. The stiffness of bonds scales as the inverse tenth power of length [114].

1.4 Biologically Synthesized Nanoparticle

Nanoparticles can be synthesized by traditional physical and chemical methods, such as laser ablation, chemical vapor deposition, ion sputtering, chemical reduction, sol-gel synthesis, etc. In addition to these methods, which we will not discuss here for the sake of brevity, are biosynthesized nanoparticles.

The synthesis of metal nanoparticles via chemical and physical methods is often expensive and may involve the toxic and hazardous chemicals. In order to mitigate these facts, green synthesis is increasingly being explored. The basic requirements for green synthesis are an ecofriendly solvent system and availability of biocompatible reducing and capping agents [81].

Metallic nanoparticles can be synthesized using biological agents, such as bacteria, yeast, fungi, plants, algae [103, 118]. The exact mechanism of biosynthesis is still unknown, except the fact that it involves proteins and/or enzymes [118].

Some bacteria developed defense mechanism to mitigate toxicity of high metal ions concentrations, such as *Pseudomonas stutzeri* that can live in soil rich in silver. Some bacteria are able to synthesize inorganic materials, such as magnetite, as in the case of the magnetotactic bacteria. A large number of bacterial species have been used to synthesize metallic NPs, some of them being listed in Table 1 [103]. Some bacteria can reduce metal ions and form them as nanoparticles inside the cells, other nanoparticles are able to synthesize them as nanoparticles both intra- and extracellularly [118].

Table 1 List of microorganisms and synthesized metal nanoparticles

Microorganisms	Products	Size (nm)	Shape
Sargassum wightii	Au	8–12	Planar
Rhodococcus sp.	Au	5–15	Spherical
Shewanella oneidensis	Au	12 ± 5	Spherical
Plectonema boryanum	Au	<10–25	Cubic
Plectonema boryanum 485	Au	10 nm–6 μm	Octahedral
Candida utilis	Au	–	–
V. luteoalbum	Au	–	–
Escherichia coli	Au	20–30	Triangles, Hexagons
Yarrowia lipolytica	Au	15	Triangles
Pseudomonas aeruginosa	Au	15–30	–
Rhodopseudomonas capsulate	Au	10–20	Spherical
Shewanella algae	Au	10–20	–
Brevibacterium casei	Au, Ag	10–50	Spherical
Trichoderma viride	Ag	5–40	Spherical
Phaenerochaete chrysosporium	Ag	50–200	Pyramidal
Bacillus licheniformis	Ag	50	–
Escherichia coli	Ag	50	–
Corynebacterium glutamicum	Ag	5–50	Irregular
Trichoderma viride	Ag	2–4	–
Ureibacillus thermosphaericus	Au	50–70	–
Bacillus cereus	Ag	4–5	Spherical
Aspergillus flavus	Ag	8.92 ± 1.61	Spherical
Aspergillus fumigatus	Ag	5–25	Spherical
Verticillium sp.	Ag	25 ± 8	Spherical
Fusarium oxysporum	Ag	5–50	Spherical
Neurospora crassa	Au, Au/Ag	32, 20–50	Spherical
Shewanella algae	Pt	5	–
Enterobacter sp.	Hg	2–5	Spherical
Shewanella sp.	Se	181 ± 40	Spherical
Escherichiacoli	Cd/Te	2.0–3.2	Spherical
Yeast	Au/Ag	9–25	Irregular polygonal
Fusarium oxysporum	Au–Ag alloy	8–14	Spherical
Pyrobaculum Islandicum	U, Tc, Cr, Co, Mn	–	Spherical
Desulfovibrio desulfuricans	Pd	50	Spherical

Reprinted from Environmental Toxicology and Pharmacology, Vol 36, issue 3, D. Nath and P. Banerjee, Green nanotechnology—A new hope for medical biology, pp. 997–1014, Copyright (2013), with permission from Elsevier [103]

Table 2 List of selected plants and synthesized nanoparticles

Plants	Nanoparticles	Shape	Size (nm)
Alfalfa plant (*Medicago sativa*)	Au and Ag	Spherical and triangular	20–40
Avena sativa	Au	Spherical	25–85
Azadirachta indica	Ag, Au	Spherical	50–100
Aloe vera	Ag	Spherical	15–15.6
Emblica officinalis	Ag and Au	Mainly spherical	10–20 and 15–25
Cinnamomum camphora	Au and Ag	Mainly spherical	55–80
Tamarind leaf extract	Au	Mainly spherical	20–40
Capsicum annum	Ag	Mainly spherical	16–40
Medicago sativa	Ti/Ni	Mainly spherical	2–6
Medicago sativa	Zn	Spherical	2–5.6

Reprinted from Environmental Toxicology and Pharmacology, Vol 36, issue 3, D. Nath and P. Banerjee, Green nanotechnology—A new hope for medical biology, pp. 997–1014, Copyright (2013), with permission from Elsevier [103]

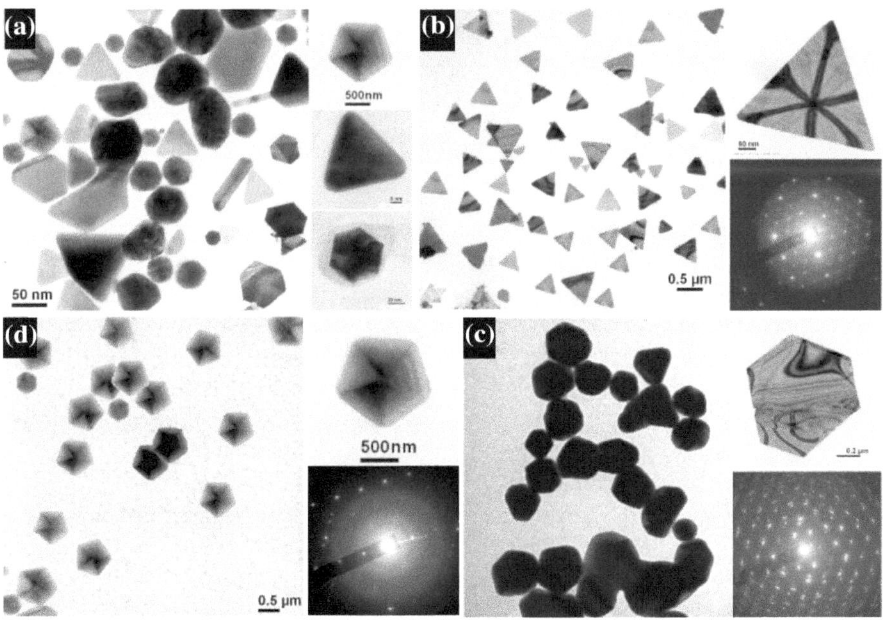

Fig. 25 TEM images of biosynthesized Au nanoparticles with different shapes under different reaction conditions (interaction of chloroauric acid and cell-free extract of the fungal strain Rhizopus oryzae); **a** mixed morphologies: nanoplates with triangle, hexagon, pentagon, star, shapes; **b** triangle; **c** hexagon; **d** pentagon. The upper and lower portions of each *right-hand* image show high-resolution single-crystalline nanoplates and their SAED patterns, respectively. Reproduced from Das et al. [33] with permission from John Wiley and Sons

A wide variety of metals nanoparticles, including semiconducting (Cd, Se, Zn, Te), alkali-earth (Cs, Sr), magnetic (Fe, Co, Ni, Mn), and noble (Au, Ag) metals and rare-earth fluorides (Pr, Gd) were synthesized using *E. Coli* [110].

The control of nanoparticles size and shape can be achieved by varying the temperature, pH, incubation time, and/or metal precursor concentration [33]. The variation in pH results in modifications in size and number of nanoparticles, while temperature changes can affect the morphology of nanoparticles and their size. The incubation time for forming nanoparticles can vary from seconds to several days [118].

Plants and plant extracts can also be used to produce metallic nanostructures, see Table 2 [103]. Plant extract is the most commonly used agent to produce metallic nanoparticles, however, some live plants are able to absorb metals from the outside and accumulate them as nanoparticles in their tissues [118].

An example of morphology and sizes of biosynthesized nanoparticles are shown in Fig. 25 [33].

1.5 Nanoparticles Applications

Nanoparticles have a multitude of applications. The antimicrobial properties of nanoparticles with different composition, such as silver [98], zinc oxide [125], palladium [1], carbon [95], have antimicrobial properties against a variety of microorganisms, such as *Staphylococcus aureus* and *Escherichia coli* [1, 129].

There are numerous applications of nanoparticles, among which are industrial applications [135], in the construction industry in cement, coatings, paints, and insulating materials [109], as electrocatalysts [9], for advanced energy conversion and storage [32, 152], in functionalized textiles [10], in food contact material [52], in cosmeceutical treatments for conditions such as photoaging, hyperpigmentation, and wrinkles [90].

Theranostic nanoparticles are nanoparticles used in individualized therapies for various diseases that combine diagnostic and therapeutic capabilities into a single agent [28]. Magnetic nanoparticles might be used as potential theranostics to detect and treat Alzheimer's disease [18] while gold nanoparticles can be used as oncotheranostics [140]. There are other various biomedical applications of nanoparticles [22, 53, 106]. Some can be used for drug delivery to specific cells [12, 26, 64, 71, 96, 151], some for imaging [29, 45, 70, 86, 134, 136], some for photothermal ablation of cancer cells [48].

While an increasing number of applications use nanoparticles and our knowledge on nanoparticle toxicity progresses, the need for special legislature regarding nanoparticles emerges [44].

As the toxicity of nanoparticles depends on their physicochemical characteristics, it is difficult to predict which nanoparticles are toxic and to what extent. The screening and prediction of toxicity of various nanoparticles should be performed before the application of the nanoparticles in various industries. Screening tools and predictive toxicological approaches are currently emerging [88, 89, 100, 105, 138].

1.6 Toxicity of Nanoparticles

When nanoparticles interact with organic matter they interact with biomolecules which form a corona and will dictate the way they interact with cells [34–36, 47, 101, 122, 123] and their biodistribution within the body and tissue accumulation within the organs [6, 68, 147].

Particle size plays an important role in the translocation of nanoparticles via the air–blood barrier in the respiratory system [17, 78, 79, 102, 132] biodistribution to organs [8, 43, 75], lung retention, and toxicity [14, 67, 137].

Oral exposure (ingestion) to some nanoparticles has shown to be associated with toxic effects and diseases [16, 23, 24, 37, 50, 51, 72]. For example, nanoparticles of various materials have been found in biopsies of patients with colon cancer [39].

Inhaled nanoparticles can enter the circulatory system and interact with blood cells [41, 42, 142]. Patients with blood diseases have been shown to have nanoparticles collected from explanted vena cava filters [40].

Epidemiologic studies show that nanoparticle exposure may lead to increased blood pressure [27]. Experiments on rats show that titanium dioxide nanoparticles promote arythmia by a direct interaction with cardiac tissue [127].

In conclusion, many nanoparticles have been found to be toxic to living organisms, their toxicity depending on various physicochemical characteristics whose relative importance is still unknown. Nanoparticles are associated to a wide range of diseases that can manifest immediately following exposure or many years later. It is impossible to extrapolate the toxicity of a nanomaterial from the properties of the bulk material, therefore a case-by-case approach is required in order to identify hazardous nanomaterials. Therefore, care must be taken while handling and using nanoparticles in applications.

References

1. Adams, C.P., Walker, K.A., Obare, S.O., Docherty, K.M.: Size-dependent antimicrobial effects of novel palladium nanoparticles. PLoS ONE **9**, e85981 (2014)
2. Agnihotri, S., Mukherji, S.: Size-controlled silver nanoparticles synthesized over the range 5–100 nm using the same protocol and their antibacterial efficacy. Rsc Adv. **4**, 3974–3983 (2014)
3. Aguado, A., Jarrold, M.F.: Melting and freezing of metal clusters. Ann. Rev. Phys. Chem. **62**(62), 151–172 (2011). and references therein
4. Aguado, A., Lopez, J.M.: Anomalous size dependence in the melting temperatures of free sodium clusters: An explanation for the calorimetry experiments. Phys. Rev. Lett. **94** (2005)

5. Akbarzadeh, A., Samiei, M., Davaran, S.: Magnetic nanoparticles: preparation, physical properties, and applications in biomedicine. Nanoscale Res. Lett. 7, 1–13 (2012)
6. Ali-Boucetta, H., Kostarelos, K.: Pharmacology of carbon nanotubes: toxicokinetics, excretion and tissue accumulation. Adv. Drug Deliv. Rev. **65**, 2111–2119 (2013)
7. Alivisatos, A.P.: Perspectives on the physical chemistry of semiconductor nanocrystals. J. Phys. Chem. **100**, 13226–13239 (1996)
8. Anderson, D.S., Patchin, E.S., Silva, R.M., Uyeminami, D.L., Sharmah, A., Guo, T., Das, G. K., Brown, J.M., Shannahan, J., Gordon, T., Chen, L.C., Pinkerton, K.E., van Winkle, L.S.: Influence of particle size on persistence and clearance of aerosolized silver nanoparticles in the rat lung. Toxicol. Sci. **144**, 366–381 (2015)
9. Anderson, R.M., Yancey, D.F., Zhang, L., Chill, S.T., Henkelman, G., Crooks, R.M.: A theoretical and experimental approach for correlating nanoparticle structure and electrocatalytic activity. Acc. Chem. Res. **48**, 1351–1357 (2015)
10. Anghel, I., Grumezescu, A.M., Andronescu, E., Anghel, A.G., Ficai, A., Saviuc, C., Grumezescu, V., Vasile, B.S., Chifiriuc, M.C.: Magnetite nanoparticles for functionalized textile dressing to prevent fungal biofilms development. Nanoscale Res. Lett. **7**, 501 (2012)
11. Anton, R.: On the reaction kinetics of Ni with amorphous carbon. Carbon **46**, 656–662 (2008)
12. Arruebo, M., Fernandez-Pacheco, R., Ibarra, M.R., Santamaria, J.: Magnetic nanoparticles for drug delivery. Nano Today **2**, 22–32 (2007)
13. Asoro, M.A., Damiano, J., Ferreira, P.J.: Size effects on the melting temperature of silver nanoparticles: in-situ TEM observations. Microsc. Microanal. **15**, 706–707 (2009)
14. Bakand, S., Hayes, A., Dechsakulthorn, F.: Nanoparticles: a review of particle toxicology following inhalation exposure. Inhal. Toxicol. **24**, 125–135 (2012)
15. Balasubramanian, K., Burghard, M.: Chemically functionalized carbon nanotubes. Small **1**, 180–192 (2005)
16. Ballestri, M., Baraldi, A., Gatti, A.M., Furci, L., Bagni, A., Loria, P., Rapana, R.M., Carulli, N., Albertazzi, A.: Liver and kidney foreign bodies granulomatosis in a patient with malocclusion, bruxism, and worn dental prostheses. Gastroenterology **121**, 1234–1238 (2001)
17. Braakhuis, H.M., Park, M.V., Gosens, I., de Jong, W.H., Cassee, F.R.: Physicochemical characteristics of nanomaterials that affect pulmonary inflammation. Part Fibre Toxicol. **11**, 18 (2014)
18. Busquets, M.A., Sabate, R., Estelrich, J.: Potential applications of magnetic particles to detect and treat Alzheimer's disease. Nanoscale Res. Lett. **9**, 538 (2014)
19. Buzea, C., Beydaghyan, G., Elliott, C., Robbie, K.: Control of power law scaling in the growth of silicon nanocolumn pseudo-regular arrays deposited by glancing angle deposition. Nanotechnology **16**, 1986–1992 (2005)
20. Buzea, C., Kaminska, K., Beydaghyan, G., Brown, T., Elliott, C., Dean, C., Robbie, K.: Thickness and density evaluation for nanostructured thin films by glancing angle deposition. J. Vac. Sci. Technol., B **23**, 2545–2552 (2005)
21. Buzea, C., Pacheco, II., Robbie, K.: Nanomaterials and nanoparticles: sources and toxicity. Biointerphases **2**, MR17-71 and references therein (2007)
22. Chatterjee, K., Sarkar, S., Jagajjanani Rao, K., Paria, S.: Core/shell nanoparticles in biomedical applications. Adv. Colloid Interf. Sci. **209**, 8–39 (2014)
23. Checchi, L., Nucci, M.C., Gatti, A.M., Mattia, D., Violante, F.S.: Sarcoidosis in a dental surgeon: a case report. J. Med. Case Rep. **4**, 259 (2010)
24. Chen, N., Wang, H., Huang, Q., Li, J., Yan, J., He, D., Fan, C., Song, H.: Long-term effects of nanoparticles on nutrition and metabolism. Small **10**, 3603–3611 (2014)
25. Cheng, B., Le, Y., Yu, J.G.: Preparation and enhanced photocatalytic activity of Ag@TiO2 core-shell nanocomposite nanowires. J. Hazard. Mater. **177**, 971–977 (2010)
26. Cho, K., Wang, X., Nie, S., Chen, Z.G., Shin, D.M.: Therapeutic nanoparticles for drug delivery in cancer. Clin. Cancer Res. Official J. Am. Associa. Cancer Res. **14**, 1310–1316 (2008)

27. Chung, M., Wang, D.D., Rizzo, A.M., Gachette, D., Delnord, M., Parambi, R., Kang, C.M., Brugge, D.: Association of PNC, BC, and PM2.5 measured at a central monitoring site with blood pressure in a predominantly near highway population. Int. J. Environ. Res. Public Health **12**, 2765–2780 (2015)
28. Clift, M.J., Dechezelles, J.F., Rothen-Rutishauser, B., Petri-Fink, A.: A biological perspective toward the interaction of theranostic nanoparticles with the bloodstream—what needs to be considered? Front Chem. **3**, 7 (2015)
29. Corr, S.A., Rakovich, Y.P., Gun'Ko, Y.K.: Multifunctional magnetic-fluorescent nanocomposites for biomedical applications. Nanoscale Res. Lett. **3**, 87–104 (2008)
30. Couchman, P.R., Jesser, W.A.: Thermodynamic theory of size dependence of melting temperature in metals. Nature **269**, 481–483 (1977)
31. Crespo, P., De La Presa, P., Marin, P., Multigner, M., Alonso, J.M., Rivero, G., Yndurain, F., Gonzalez-Calbet, J.M., Hernando, A.: Magnetism in nanoparticles: tuning properties with coatings. J. Phys. Condensed Matter. **25** (2013)
32. Dai, L., Chang, D.W., Baek, J.B., Lu, W.: Carbon nanomaterials for advanced energy conversion and storage. Small **8**, 1130–1166 (2012)
33. Das, S.K., Das, A.R., Guha, A.K.: Microbial synthesis of multishaped gold nanostructures. Small **6**, 1012–1021 (2010)
34. Docter, D., Westmeier, D., Markiewicz, M., Stolte, S., Knauer, S.K., Stauber, R.H.: The nanoparticle biomolecule corona: lessons learned—challenge accepted? Chem. Soc. Rev. **44**, 6094–6121 (2015)
35. Fleischer, C.C., Payne, C.K.: Nanoparticle-cell interactions: molecular structure of the protein corona and cellular outcomes. Acc. Chem. Res. **47**, 2651–2659 (2014)
36. Foroozandeh, P., Aziz, A.A.: Merging worlds of nanomaterials and biological environment: factors governing protein corona formation on nanoparticles and its biological consequences. Nanoscale Res. Lett. **10**, 221 (2015)
37. Gaillet, S., Rouanet, J.M.: Silver nanoparticles: their potential toxic effects after oral exposure and underlying mechanisms–a review. Food Chem. Toxicol. **77**, 58–63 (2015)
38. Gajewicz, A., Schaeublin, N., Rasulev, B., Hussain, S., Leszczynska, D., Puzyn, T., Leszczynski, J.: Towards understanding mechanisms governing cytotoxicity of metal oxides nanoparticles: hints from nano-QSAR studies. Nanotoxicology **9**, 313–325 (2015)
39. Gatti, A.M.: Biocompatibility of micro- and nano-particles in the colon. Part II. Biomater. **25**, 385–392 (2004)
40. Gatti, A.M., Montanari, S.: Retrieval analysis of clinical explanted vena cava filters. J. Biomed. Mater. Res. B Appl. Biomater. **77**, 307–314 (2006)
41. Gatti, A.M., Montanari, S., Monari, E., Gambarelli, A., Capitani, F., Parisini, B.: Detection of micro- and nano-sized biocompatible particles in the blood. J. Mater. Sci. Mater. Med. **15**, 469–472 (2004)
42. Gatti, A.M., Montanari, S., Gambarelli, A., Capitani, F., Salvatori, R.: In-vivo short- and long-term evaluation of the interaction material-blood. J. Mater. Sci. Mater. Med. **16**, 1213–1219 (2005)
43. Geiser, M., Kreyling, W.G.: Deposition and biokinetics of inhaled nanoparticles. Part Fibre Toxicol. **7**, 2 (2010)
44. Godwin, H., Nameth, C., Avery, D., Bergeson, L.L., Bernard, D., Beryt, E., Boyes, W., Brown, S., Clippinger, A.J., Cohen, Y., Doa, M., Hendren, C.O., Holden, P., Houck, K., Kane, A.B., Klaessig, F., Kodas, T., Landsiedel, R., Lynch, I., Malloy, T., Miller, M.B., Muller, J., Oberdorster, G., Petersen, E.J., Pleus, R.C., Sayre, P., Stone, V., Sullivan, K.M., Tentschert, J., Wallis, P., Nel, A.E.: Nanomaterial categorization for assessing risk potential to facilitate regulatory decision-making. ACS Nano **9**, 3409–3417 (2015)
45. Gong, M., Yang, H., Zhang, S., Yang, Y., Zhang, D., Qi, Y., Zou, L.: Superparamagnetic core/shell GoldMag nanoparticles: size-, concentration- and time-dependent cellular nanotoxicity on human umbilical vein endothelial cells and the suitable conditions for magnetic resonance imaging. J. Nanobiotechnol. **13**, 24 (2015)

46. Greget, R., Nealon, G.L., Vileno, B., Turek, P., Meny, C., Ott, F., Derory, A., Voirin, E., Riviere, E., Rogalev, A., Wilhelm, F., Joly, L., Knafo, W., Ballon, G., Terazzi, E., Kappler, J.P., Donnio, B., Gallani, J.L.: Magnetic properties of gold nanoparticles: a room-temperature quantum effect. ChemPhysChem **13**, 3092–3097 (2012)
47. Grillo, R., Rosa, A.H., Fraceto, L.F.: Engineered nanoparticles and organic matter: a review of the state-of-the-art. Chemosphere **119**, 608–619 (2015)
48. Guo, Y., Zhang, Z., Kim, D.H., Li, W., Nicolai, J., Procissi, D., Huan, Y., Han, G., Omary, R.A., Larson, A.C.: Photothermal ablation of pancreatic cancer cells with hybrid iron-oxide core gold-shell nanoparticles. Int. J. Nanomedicine **8**, 3437–3446 (2013)
49. Haberland, H., Hippler, T., Donges, J., Kostko, O., Schmidt, M., von Issendorff, B.: Melting of sodium clusters: where do the magic numbers come from? Phys. Rev. Lett. **94**, 035701 (2005)
50. Hadrup, N., Lam, H.R.: Oral toxicity of silver ions, silver nanoparticles and colloidal silver—a review. Regul. Toxicol. Pharmacol. **68**, 1–7 (2014)
51. Hadrup, N., Sharma, A.K., Poulsen, M., Nielsen, E.: Toxicological risk assessment of elemental gold following oral exposure to sheets and nanoparticles—A review. Regul. Toxicol. Pharmacol. **72**, 216–221 (2015)
52. Hannon, J.C., Kerry, J., Cruz-Romero, M., Morris, M., Cummins, E.: Advances and challenges for the use of engineered nanoparticles in food contact materials. Trends Food Sci. Technol. **43**, 43–62 (2015)
53. Hofmann-Amtenbrink, M., Grainger, D.W., Hofmann, H.: Nanoparticles in medicine: current challenges facing inorganic nanoparticle toxicity assessments and standardizations. http://www.ncbi.nlm.nih.gov/pubmed/26051651 (2015). Accessed 7 Nov 2015
54. Hori, H., Teranishi, T., Nakae, Y., Seino, Y., Miyake, M., Yamada, S.: Anomalous magnetic polarization effect of Pd and Au nano-particles. Phys. Lett. A **263**, 406–410 (1999)
55. Hori, H., Yamamoto, Y., Iwamoto, T., Miura, T., Teranishi, T., Miyake, M.: Diameter dependence of ferromagnetic spin moment in Au nanocrystals. Phys. Rev. B, **69**, (2004)
56. Hu, Y., Peng, K.Q., Liu, L., Qiao, Z., Huang, X., Wu, X.L., Meng, X.M., Lee, S.T.: Continuous-flow mass production of silicon nanowires via substrate-enhanced metal-catalyzed electroless etching of silicon with dissolved oxygen as an oxidant. Sci. Reports, **4** (2014)
57. Huang, X., Teng, X., Chen, D., Tang, F., He, J.: The effect of the shape of mesoporous silica nanoparticles on cellular uptake and cell function. Biomaterials **31**, 438–448 (2010)
58. Huang, X.H., Jain, P.K., El-Sayed, I.H., El-Sayed, M.A.: Gold nanoparticles: interesting optical properties and recent applications in cancer diagnostic and therapy. Nanomedicine **2**, 681–693 (2007). and references therein
59. Huber, D.L.: Synthesis, properties, and applications of iron nanoparticles. Small **1**, 482–501 (2005)
60. Hussein, M.Z., Azmin, W., Mustafa, M., Yahaya, A.H.: Bacillus cereus as a biotemplating agent for the synthesis of zinc oxide with raspberry- and plate-like structures. J. Inorg. Biochem. **103**, 1145–1150 (2009)
61. Issa, B., Obaidat, I.M., Albiss, B.A., Haik, Y.: Magnetic nanoparticles: surface effects and properties related to biomedicine applications. Int. J. Mol. Sci. **14**, 21266–21305 (2013)
62. Jeong, U., Teng, X.W., Wang, Y., Yang, H., Xia, Y.N.: Superparamagnetic colloids: controlled synthesis and niche applications. Adv. Mater. **19**, 33–60 (2007)
63. Jin, M., Zhang, H., Wang, J., Zhong, X., Lu, N., Li, Z., Xie, Z., Kim, M.J., Xia, Y.: Copper can still be epitaxially deposited on palladium nanocrystals to generate core-shell nanocubes despite their large lattice mismatch. ACS Nano **6**, 2566–2573 (2012)
64. Keck, C.M., Muller, R.H.: Nanotoxicological classification system (NCS)—a guide for the risk-benefit assessment of nanoparticulate drug delivery systems. Eur. J. Pharm. Biopharm. **84**, 445–448 (2013)
65. Kedia, A., Kumar, P.S.: Precursor-driven nucleation and growth kinetics of gold nanostars. J. Phys. Chem. C **116**, 1679–1686 (2012)

66. Kedia, A., Kumar, H., Kumar, P.S.: Tweaking anisotropic gold nanostars: covariant control of a polymer- solvent mixture complex. Rsc Adv. **5**, 5205–5212 (2015)
67. Keller, J., Wohlleben, W., Ma-Hock, L., Strauss, V., Groters, S., Kuttler, K., Wiench, K., Herden, C., Oberdorster, G., van Ravenzwaay, B., Landsiedel, R.: Time course of lung retention and toxicity of inhaled particles: short-term exposure to nano-Ceria. Arch. Toxicol. **88**, 2033–2059 (2014)
68. Khlebtsov, N., Dykman, L.: Biodistribution and toxicity of engineered gold nanoparticles: a review of in vitro and in vivo studies. Chem. Soc. Rev. **40**, 1647–1671 (2011)
69. Khlebtsov, N.G., Dykman, L.A.: Optical properties and biomedical applications of plasmonic nanoparticles. J. Quant. Spectrosc. Radiat. Transfer **111**, 1–35 (2010)
70. Kiessling, F., Mertens, M.E., Grimm, J., Lammers, T.: Nanoparticles for imaging: top or flop? Radiology **273**, 10–28 (2014)
71. Kim, J.E., Shin, J.Y., Cho, M.H.: Magnetic nanoparticles: an update of application for drug delivery and possible toxic effects. Arch. Toxicol. **86**, 685–700 (2012)
72. Kim, Y.S., Kim, J.S., Cho, H.S., Rha, D.S., Kim, J.M., Park, J.D., Choi, B.S., Lim, R., Chang, H.K., Chung, Y.H., Kwon, I.H., Jeong, J., Han, B.S., Yu, I.J.: Twenty-eight-day oral toxicity, genotoxicity, and gender-related tissue distribution of silver nanoparticles in Sprague-Dawley rats. Inhal. Toxicol. **20**, 575–583 (2008)
73. Klabunde, K.J., Stark, J., Koper, O., Mohs, C., Park, D.G., Decker, S., Jiang, Y., Lagadic, I., Zhang, D.J.: Nanocrystals as stoichiometric reagents with unique surface chemistry. J. Phys. Chem. **100**, 12142–12153 (1996)
74. Kolhatkar, A.G., Jamison, A.C., Litvinov, D., Willson, R.C., Lee, T.R.: Tuning the magnetic properties of nanoparticles. Int. J. Mol. Sci. **14**, 15977–16009 (2013)
75. Konduru, N., Keller, J., Ma-Hock, L., Groters, S., Landsiedel, R., Donaghey, T.C., Brain, J. D., Wohlleben, W., Molina, R.M.: Biokinetics and effects of barium sulfate nanoparticles. Part Fibre Toxicol. **11**, 55 (2014)
76. Korth, B.D., Keng, P., Shim, I., Bowles, S.E., Tang, C., Kowalewski, T., Nebesny, K.W., Pyun, J.: Polymer-coated ferromagnetic colloids from well-defined macromolecular surfactants and assembly into nanoparticle chains. J. Am. Chem. Soc. **128**, 6562–6563 (2006)
77. Krahne, R., Morello, G., Figuerola, A., George, C., Deka, S., Manna, L.: Physical properties of elongated inorganic nanoparticles. Phys. Reports Rev. Sect. Phys. Lett. **501**, 75–221 (2011)
78. Kreyling, W.G., Semmler-Behnke, M., Takenaka, S., Moller, W.: Differences in the biokinetics of inhaled nano- versus micrometer-sized particles. Acc. Chem. Res. **46**, 714–722 (2013)
79. Kreyling, W.G., Hirn, S., Moller, W., Schleh, C., Wenk, A., Celik, G., Lipka, J., Schaffler, M., Haberl, N., Johnston, B.D., Sperling, R., Schmid, G., Simon, U., Parak, W.J., Semmler-Behnke, M.: Air-blood barrier translocation of tracheally instilled gold nanoparticles inversely depends on particle size. ACS Nano **8**, 222–233 (2014)
80. Krishna, K.S., Tarakeshwar, P., Mujica, V., Kumar, C.: Chemically induced magnetism in atomically precise gold clusters. Small **10**, 907–911 (2014)
81. Kulkarni, R.R., Shaiwale, N.S., Deobagkar, D.N., Deobagkar, D.D.: Synthesis and extracellular accumulation of silver nanoparticles by employing radiation-resistant Deinococcus radiodurans, their characterization, and determination of bioactivity. Int. J. Nanomed. **10**, 963–974 (2015)
82. Lai, S.L., Guo, J.Y., Petrova, V., Ramanath, G., Allen, L.H.: Size-dependent melting properties of small tin particles: Nanocalorimetric measurements. Phys. Rev. Lett. **77**, 99–102 (1996)
83. Leonard, S.S., Cohen, G.M., Kenyon, A.J., Schwegler-Berry, D., Fix, N.R., Bangsaruntip, S., Roberts, J.R.: Generation of reactive oxygen species from silicon nanowires. Environ. Health Insights **8**, 21–29 (2014)
84. Li, C.Y., Wu, C.M., Karna, S.K., Wang, C.W., Hsu, D., Wang, C.J., Li, W.H.: Intrinsic magnetic moments of gold nanoparticles. Phys. Rev. B, **83** (2011)

85. Li, J., Hietala, S., Tian, X.: BaTiO3 supercages: unusual oriented nanoparticle aggregation and continuous ordering transition in morphology. ACS Nano **9**, 496–502 (2015)
86. Li, M., Kim, H.S., Tian, L., Yu, M.K., Jon, S., Moon, W.K.: Comparison of two ultrasmall superparamagnetic iron oxides on cytotoxicity and MR imaging of tumors. Theranostics **2**, 76–85 (2012)
87. Liu, L., Yoo, S.H., Lee, S.A., Park, S.: Wet-chemical synthesis of palladium nanosprings. Nano Lett. **11**, 3979–3982 (2011)
88. Liu, R., Rallo, R., George, S., Ji, Z., Nair, S., Nel, A.E., Cohen, Y.: Classification NanoSAR development for cytotoxicity of metal oxide nanoparticles. Small **7**, 1118–1126 (2011)
89. Liu, R., Rallo, R., Weissleder, R., Tassa, C., Shaw, S., Cohen, Y.: Nano-SAR development for bioactivity of nanoparticles with considerations of decision boundaries. Small **9**, 1842–1852 (2013)
90. Lohani, A., Verma, A., Joshi, H., Yadav, N., Karki, N.: Nanotechnology-based cosmeceuticals. ISRN Dermatol **2014**, 843687 (2014)
91. Lue, J.T.: A review of characterization and physical property studies of metallic nanoparticles. J. Phys. Chem. Solids **62**, 1599–1612 (2001)
92. Luo, L.L., Wu, J.S., Luo, J.Y., Huang, J.X., Dravid, V.P.: Dynamics of electrochemical lithiation/delithiation of graphene-encapsulated silicon nanoparticles studied by In-situ TEM. Sci. Reports, **4** (2014)
93. Ma-Hock, L., Strauss, V., Treumann, S., Kuttler, K., Wohlleben, W., Hofmann, T., Groters, S., Wiench, K., van Ravenzwaay, B., Landsiedel, R.: Comparative inhalation toxicity of multi-wall carbon nanotubes, graphene, graphite nanoplatelets and low surface carbon black. Part Fibre Toxicol. **10**, 23 (2013)
94. Maitra, U., Das, B., Kumar, N., Sundaresan, A., Rao, C.N.R.: Ferromagnetism exhibited by nanoparticles of noble metals. ChemPhysChem **12**, 2322–2327 (2011)
95. Maleki Dizaj, S., Mennati, A., Jafari, S., Khezri, K., Adibkia, K., Antimicrobial activity of carbon-based nanoparticles. Adv. Pharm. Bull. **5**, 19–23 (2015)
96. Mandal, B., Bhattacharjee, H., Mittal, N., Sah, H., Balabathula, P., Thoma, L.A., Wood, G. C.: Core-shell-type lipid-polymer hybrid nanoparticles as a drug delivery platform. Nanomed. Nanotechnol. Biol. Med. **9**, 474–491 (2013)
97. Mark, A.G., Gibbs, J.G., Lee, T.C., Fischer, P.: Hybrid nanocolloids with programmed three-dimensional shape and material composition. Nat. Mater. **12**, 802–807 (2013)
98. Marx, D.E., Barillo, D.J.: Silver in medicine: the basic science. Burns J. Int. Soc. Burn Injuries **40**(Suppl 1), S9–S18 (2014)
99. Mattox, T.M., Ye, X.C., Manthiram, K., Schuck, P.J., Alivisatos, A.P., Urban, J.J.: Chemical control of plasmons in metal chalcogenide and metal oxide nanostructures. Adv. Mater. **27**, 5830–5837 (2015)
100. Meindl, C., Kueznik, T., Bosch, M., Roblegg, E., Frohlich, E.: Intracellular calcium levels as screening tool for nanoparticle toxicity. J. Appl. Toxicol. **35**, 1150–1159 (2015)
101. Monopoli, M.P., Aberg, C., Salvati, A., Dawson, K.A.: Biomolecular coronas provide the biological identity of nanosized materials. Nat. Nanotechnol. **7**, 779–786 (2012)
102. Nakane, H.: Translocation of particles deposited in the respiratory system: a systematic review and statistical analysis. Environ. Health Prev. Med. **17**, 263–274 (2012)
103. Nath, D., Banerjee, P.: Green nanotechnology—a new hope for medical biology. Environ. Toxicol. Pharmacol. **36**, 997–1014 (2013)
104. Nealon, G.L., Donnio, B., Greget, R., Kappler, J.P., Terazzi, E., Gallani, J.L.: Magnetism in gold nanoparticles. Nanoscale **4**, 5244–5258 (2012)
105. Nel, A., Xia, T., Meng, H., Wang, X., Lin, S., Ji, Z., Zhang, H.: Nanomaterial toxicity testing in the 21st century: use of a predictive toxicological approach and high-throughput screening. Acc. Chem. Res. **46**, 607–621 (2013)
106. Nie, S., Xing, Y., Kim, G.J., Simons, J.W.: Nanotechnology applications in cancer. Ann. Rev. Biomed. Eng. **9**, 257–288 (2007)
107. Noya, E.G., Doye, J.P.K., Wales, D.J., Aguado, A.: Geometric magic numbers of sodium clusters: interpretation of the melting behaviour. Euro. Phys. J. D **43**, 57–60 (2007)

108. Ojha, U., Steenbergen, K.G., Gaston, N.: Al20 + does melt, albeit above the bulk melting temperature of aluminium. Phys. Chem. Chem. Phys. **17**, 3741–3748 (2015)
109. Pacheco Blandino, I., Vanner, R., Buzea, C.: Toxicity of nanoparticles. In: Pacheco-Torgal, F., Jalali, S., Fucic, A. (eds.) Toxicity of Building Materials. Woodhead (2012)
110. Park, T.J., Lee, S.Y., Heo, N.S., Seo, T.S.: In vivo synthesis of diverse metal nanoparticles by recombinant escherichia coli. Angewandte Chemie-International Edition **49**, 7019–7024 (2010)
111. Peng, Z.M., Yang, H.: Designer platinum nanoparticles: control of shape, composition in alloy, nanostructure and electrocatalytic property. Nano Today **4**, 143–164 (2009)
112. Podila, R., Brown, J.M.: Toxicity of engineered nanomaterials: a physicochemical perspective. J. Biochem. Mol. Toxicol. **27**, 50–55 (2013). and references therein
113. Potocnik, J.: Commission recommendation of 18 October 2011 on the definition of nanomaterial (Text with EEA relevance) (2011/696/EU). In: Commission, T.E. (ed.) Official Journal of the European Union (2011)
114. Pushpa, R., Waghmare, U., Narasimhan, S.: Bond stiffening in small nanoclusters and its consequences for mechanical and thermal properties. Phys. Rev. B **77**, 045427 (2008)
115. Pyfer, K.L., Kafader, J.O., Yalamanchali, A., Jarrold, M.F.: Melting of size-selected gallium clusters with 60–183 atoms. J. Phys. Chem. A **118**, 4900–4906 (2014)
116. Qi, W.H.: Size effect on melting temperature of nanosolids. Phys. B-Condensed Matter. **368**, 46–50 (2005)
117. Qiu, Y., Liu, Y., Wang, L., Xu, L., Bai, R., Ji, Y., Wu, X., Zhao, Y., Li, Y., Chen, C.: Surface chemistry and aspect ratio mediated cellular uptake of Au nanorods. Biomaterials **31**, 7606–7619 (2010)
118. Quester, K., Avalos-Borja, M., Castro-Longoria, E.: Biosynthesis and microscopic study of metallic nanoparticles. Micron **54–55**, 1–27 (2013). and references therein
119. Reich, S., Leitus, G., Feldman, Y.: Observation of magnetism in Au thin films. Appl. Phys. Lett. **88** (2006)
120. Robbie, K., Sit, J.C., Brett, M.J.: Advanced techniques for glancing angle deposition. J. Vacuum Sci. Technol. B **16**, 1115–1122 (1998)
121. Robbie, K., Beydaghyan, G., Brown, T., Dean, C., Adams, J., Buzea, C.: Ultrahigh vacuum glancing angle deposition system for thin films with controlled three-dimensional nanoscale structure. Rev. Sci. Instrum. **75**, 1089–1097 (2004)
122. Roy, R., Kumar, S., Tripathi, A., Das, M., Dwivedi, P.D.: Interactive threats of nanoparticles to the biological system. Immunol. Lett. **158**, 79–87 (2014)
123. Saha, K., Bajaj, A., Duncan, B., Rotello, V.M.: Beauty is skin deep: a surface monolayer perspective on nanoparticle interactions with cells and biomacromolecules. Small **7**, 1903–1918 (2011)
124. Sakamoto, Y., Oba, Y., Maki, H., Suda, M., Einaga, Y., Sato, T., Mizumaki, M., Kawamura, N., Suzuki, M.: Ferromagnetism of Pt nanoparticles induced by surface chemisorption. Phys. Rev. B **83** (2011)
125. Saliani, M., Jalal, R., Kafshdare Goharshadi, E.: Effects of pH and temperature on antibacterial activity of zinc oxide nanofluid against escherichia coli O157: H7 and staphylococcus aureus. Jundishapur J. Microbiol. **8**, e17115 (2015)
126. Sato, R., Ishikawa, S., Sato, H., Sato, T.: Magnetic order of Au nanoparticle with clean surface. J. Mag. Mag. Mater. **393**, 209–212 (2015). and references therein
127. Savi, M., Rossi, S., Bocchi, L., Gennaccaro, L., Cacciani, F., Perotti, A., Amidani, D., Alinovi, R., Goldoni, M., Aliatis, I., Lottici, P.P., Bersani, D., Campanini, M., Pinelli, S., Petyx, M., Frati, C., Gervasi, A., Urbanek, K., Quaini, F., Buschini, A., Stilli, D., Rivetti, C., Macchi, E., Mutti, A., Miragoli, M., Zaniboni, M.: Titanium dioxide nanoparticles promote arrhythmias via a direct interaction with rat cardiac tissue. Part Fibre Toxicol. **11**, 63 (2014)
128. Scholes, G.D.: Controlling the optical properties of inorganic nanoparticles. Adv. Funct. Mater. **18**, 1157–1172 (2008)
129. Seil, J.T., Webster, T.J.: Antimicrobial applications of nanotechnology: methods and literature. Int. J. Nanomed. **7**, 2767–2781 (2012)

130. Shinohara, T., Sato, T., Taniyama, T.: Surface ferromagnetism of Pd fine particles. Phys. Rev. Lett. **91** (2003)
131. Shvartsburg, A.A., Jarrold, M.F.: Solid clusters above the bulk melting point. Phys. Rev. Lett. **85**, 2530–2532 (2000)
132. Silva, R.M., Doudrick, K., Franzi, L.M., Teesy, C., Anderson, D.S., Wu, Z., Mitra, S., Vu, V., Dutrow, G., Evans, J.E., Westerhoff, P., van Winkle, L.S., Raabe, O.G., Pinkerton, K.E.: Instillation versus inhalation of multiwalled carbon nanotubes: exposure-related health effects, clearance, and the role of particle characteristics. ACS Nano **8**, 8911–8931 (2014)
133. Smith, A.M., Nie, S.: Chemical analysis and cellular imaging with quantum dots. Analyst **129**, 672–677 (2004)
134. Smith, A.M., Duan, H., Mohs, A.M., Nie, S.: Bioconjugated quantum dots for in vivo molecular and cellular imaging. Adv. Drug Deliv. Rev. **60**, 1226–1240 (2008)
135. Stark, W.J., Stoessel, P.R., Wohlleben, W., Hafner, A.: Industrial applications of nanoparticles. Chem. Soc. Rev. **44**, 5793–5805 (2015)
136. Sun, X., Cai, W., Chen, X.: Positron emission tomography imaging using radiolabeled inorganic nanomaterials. Acc. Chem. Res. **48**, 286–294 (2015)
137. Theodorou, I.G., Ryan, M.P., Tetley, T.D., Porter, A.E.: Inhalation of silver nanomaterials–seeing the risks. Int. J. Mol. Sci. **15**, 23936–23974 (2014)
138. Thomas, C.R., George, S., Horst, A.M., Ji, Z., Miller, R.J., Peralta-Videa, J.R., Xia, T., Pokhrel, S., Madler, L., Gardea-Torresdey, J.L., Holden, P.A., Keller, A.A., Lenihan, H.S., Nel, A.E., Zink, J.I.: Nanomaterials in the environment: from materials to high-throughput screening to organisms. ACS Nano **5**, 13–20 (2011)
139. Tottori, S., Zhang, L., Qiu, F., Krawczyk, K.K., Franco-Obregon, A., Nelson, B.J.: Magnetic helical micromachines: fabrication, controlled swimming, and cargo transport. Adv. Mater. **24**, 811–816 (2012)
140. Trouiller, A.J., Hebie, S., el Bahhaj, F., Napporn, T.W., Bertrand, P.: Chemistry for oncotheranostic gold nanoparticles. Eur. J. Med. Chem. **99**, 92–112 (2015)
141. Volokitin, Y., Sinzig, J., Dejongh, L.J., Schmid, G., Vargaftik, M.N., Moiseev, II.: Quantum-size effects in the thermodynamic properties of metallic nanoparticles. Nature **384**, 621–623 (1996)
142. Wang, B., He, X., Zhang, Z., Zhao, Y., Feng, W.: Metabolism of nanomaterials in vivo: blood circulation and organ clearance. Acc. Chem. Res. **46**, 761–769 (2013)
143. Xia, Y.S., Tang, Z.Y.: Monodisperse hollow supraparticles via selective oxidation. Adv. Funct. Mater. **22**, 2585–2593 (2012)
144. Xie, J.P., Lee, J.Y., Wang, D.I.C., Ting, Y.P.: Identification of active biomolecules in the high-yield synthesis of single-crystalline gold nanoplates in algal solutions. Small **3**, 672–682 (2007)
145. Yamamoto, Y., Miura, T., Suzuki, M., Kawamura, N., Miyagawa, H., Nakamura, T., Kobayashi, K., Teranishi, T., Hori, H.: Direct observation of ferromagnetic spin polarization in gold nanoparticles. Phys. Rev. Lett. **93** (2004)
146. Yang, J.Y., Sun, Y., He, L., Xiong, C.M., Dou, R.F., Nie, J.C.: Size-dependent magnetic moments in ultrafine diamagnetic systems. J. Appl. Phys. **109** (2011)
147. Yang, S.T., Guo, W., Lin, Y., Deng, X.Y., Wang, H.F., Sun, H.F., Liu, Y.F., Wang, X., Wang, W., Chen, M., Huang, Y.P., Sun, Y.P.: Biodistribution of pristine single-walled carbon nanotubes in vivo. J. Phys. Chem. C **111**, 17761–17764 (2007)
148. Yang, Y., Zheng, Y., Cao, W., Titov, A., Hyvonen, J., Mandersjesse, R., Xue, J., Holloway, P.H., Qian, L.: High-efficiency light-emitting devices based on quantum dots with tailored nanostructures. Nat. Photon. **9**, 259–266 (2015)
149. Ye, L., Yong, K.-T., Liu, L., Roy, I., Hu, R., Zhu, J., Cai, H., Law, W.-C., Liu, J., Wang, K., Liu, J., Liu, Y., Hu, Y., Zhang, X., Swihart, M.T., Prasad, P.N.: A pilot study in non-human primates shows no adverse response to intravenous injection of quantum dots. Nat. Nano **7**, 453–458 (2012)
150. Yu, X., Zhan, Z.: The effects of the size of nanocrystalline materials on their thermodynamic and mechanical properties. Nano. Res. Lett. **9**, 516 (2014)

151. Zensi, A., Begley, D., Pontikis, C., Legros, C., Mihoreanu, L., Wagner, S., Buchel, C., von Briesen, H., Kreuter, J.: Albumin nanoparticles targeted with Apo E enter the CNS by transcytosis and are delivered to neurones. J. Controlled Release Official J. Controlled Release Soc. **137**, 78–86 (2009)
152. Zhang, Q., Huang, J.Q., Qian, W.Z., Zhang, Y.Y., Wei, F.: The road for nanomaterials industry: a review of carbon nanotube production, post-treatment, and bulk applications for composites and energy storage. Small **9**, 1237–1265 (2013)

A Brief Manifestation of Nanotechnology

Sharda Sundaram Sanjay and Avinash C. Pandey

Abstract Nanotechnology is basically focused on the fabrication of nanomaterials based on the manipulation, control, and integration of atoms and molecules at nanometer scale, due to which there develops sudden change in the size-dependent properties and functions. We have to integrate chemistry, physics and biology to form materials, structures, components, devices, and systems at the nanoscale level. In the metric system, prefix "nano-" refers to one-billionth (0.000 000 001 = 10^{-9}) of the base unit. In the nanoscale, at least one of the particle's dimensions (height, width, or depth) should be at less than 100 nm. At this level, the chemical reactivity changes dramatically due to the reduction of particle size. This change occurs as a function of the structure and the density of electrons in the outermost electronic energy levels. Along with the physical properties such as optical, electrical, and thermal properties, magnetic characteristics may also change which in-turn depend on the distribution of electrons in the outermost energy levels, leads to the novel optical, electrical, magnetic behaviors and changes in the surface dependent properties. Because of the increase in the surface-to-volume ratio at the nanoscale level the properties of the material become strongly dependent (controllable) on the surface of the materials. The nanomaterials may be classified in number of ways, viz., based on: (i) dimensionality (ii) surface morphology (iii) crystalline forms (iv) chemical nature (v) chemical composition (vi) magnetic behavior (vii) functionalization or (viii) applications. But none of these classifications can be considered as absolute one. They are usually composites or hybrid in nature because in majority of cases organic compounds are used to stabilize them through capping or functionalization during synthesis. Magnetic nanomaterials have giant spins, which

S.S. Sanjay (✉)
Chemistry Department, Ewing Christian College, Allahabad, India
e-mail: sharda100@redifmail.com

A.C. Pandey
Nanotechnology Application Centre, University of Allahabad, Allahabad, India

A.K. Shukla (ed.), *EMR/ESR/EPR Spectroscopy for Characterization of Nanomaterials*, Advanced Structured Materials 62,
DOI 10.1007/978-81-322-3655-9_2

may be described as a single magnetic domain having uniaxial anisotropy, their EMR signals can be exploited for various biomedical and many other applications. Number of biomedical applications of nanoparticles is becoming possible because of their specific physicochemical properties and controllable dimensions which range from a few nanometers to nearly tens of nanometer, due to which they come in the size range smaller than that of a plant or animal cells having submicron size domain.

Keywords Nanomaterials · Classification · Dimensionality · Surface morphology · Chemical nature

1 Introduction

It will not be an exaggeration to call Nobel Prize winner physicist of the year 1965, Sir Richard Feynman as "Father of nanotechnology," as he came up with his historical thought of nano by uttering these words, "there is plenty of room at the bottom" in his historic lecture at a conference held by American Physical Society in 1959, though the term nanotechnology was first given by Norio Taniguchi [1] at the University of Tokyo in 1974. Feynman introduced the idea of manipulating the very small, even down to the level of individual atoms for technological purposes. Within the span of a half century nanotechnology industry has taken over. Nanotechnology generally means building things from the bottom level, with atomic precision. Today, nanotechnology is meant for the science of making, synthesizing, designing, producing things which have at least at one dimensions of 100 nm or less and using it for various devices. It means that purposeful engineering of matter by manipulation, control, and integration of atoms and molecules at scales of 100 nm or less forms the basis of nanotechnology to achieve size-dependent properties and functions. This integrates chemistry, physics and biology to form materials, structures, components, devices, and systems to play at nanoscale level especially for industrial and commercial objectives.

Due to their peculiar size and physicochemical properties, nanomaterials offers major advantages. The development of unique nanoscale structures has revolutionized the industry. In the realm of nanotechnology the scientists, chemists, physicist and engineers have started working at the molecular and cellular levels to make important advances (Fig. 1) in various fields such as fabrication, modeling, life sciences, medicine, magnets and healthcare, electronics, smart materials and sensors, nanoscale biostructures, energy capture and storage devices, and consumer products.

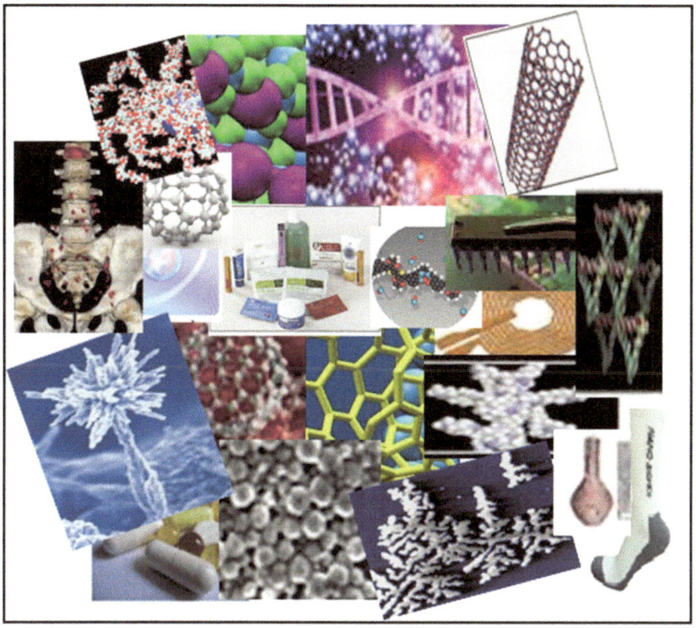

Fig. 1 Materials with nanotechnology

2 Classification

Each and every material is made up of peculiar arrangement of particular atoms in a specific way which defines its properties and behavior. Nanomaterials can be prepared by cutting, chipping, pounding, extruding, ball milling, precipitating and many more top–down and bottom–up methods. The main difference in the structure of nanomaterials arises on how they are prepared. Their formation requires knowledge of new types of synthetic methods and new understanding of the formation of materials at nanoscale. In the case of nanomaterials, the atoms align themselves in space in such a different way which is not found in the bulk form of the same material and there is significant variation in their properties also. According to the difference in certain criteria, nanomaterials may be classified in number of ways. Classifications based on some important criterions are given below

2.1 Classification Based on Dimensionality

Nanomaterials can be created with various modulation dimensionalities. Siegel [2] classified these nanostructured materials accordingly into four categories based on their dimensionality as

(i) Zero dimensional (0D)—Having three dimensions in the nanometeric range (by a dimensionality limited to 50 nm), e.g., atomic clusters, filaments and cluster assemblies, fullerenes, etc.
(ii) One dimensional (1D)—Having two dimensions in the nanometeric range, e.g., multilayers, nanotubes, nanorodes, nanowires, nanofibers manifest more advanced and promising properties as being the 1D quantum wires nanoscopic in diameter but microscopic in length.
(iii) Two dimensional (2D)—These have one dimension in nanometeric range, e.g., ultrafine-grained over-layers or buried layers, heterostructures, and nanodisks.
(iv) Three dimensional (3D)—Such nanophase materials which consist of equiaxed nanometer sized grains. For example, thin films with atomic-scale porosity, colloidal and different nanoparticles having various structures and morphologies.

Classification based on dimensionality is pictorially represented in Fig. 2.

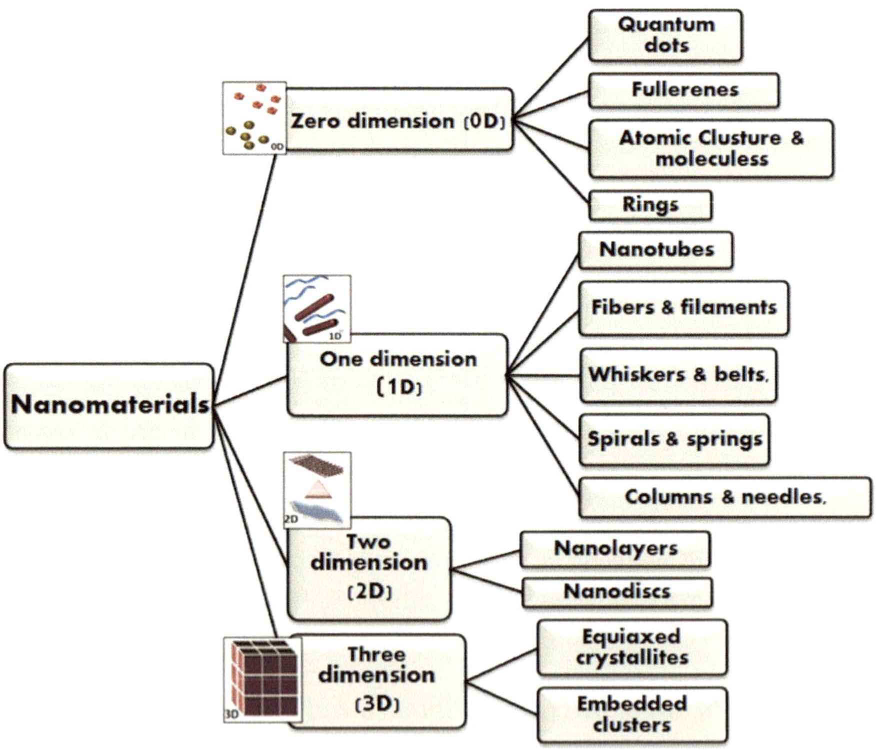

Fig. 2 Classification of nanomaterials based on dimensionality

2.2 Classification Based on *Surface Morphology*

As a reflection of templation or mode of synthesis different types of morphologies or shapes arise spontaneously. The properties and usage of a particular nanoparticles depends on its morphology provides the basic key for their mode of utilization in several advancing technologies. There are nanospheres that are spherical, nanoreefs, nanoboxes, nanoclusters, nanotubes, nanoflowers, etc. They may be classified on the basis of morphology considering their flatness, sphericity and aspect ratio. Aspect ratio is attributed to the proportional relationship between the widths of a structure with its height. Nanoparticles can be classified according to their aspect ratios also as

(i) Nanoparticles in which diameter varies with length have generally high aspect ratio. For example, nanotubes, nanorods, and nanowires having various shapes as helices, zigzags, nanobelts, etc.
(ii) Collections of many particles that exist as powders, suspension, or colloids have generally small-aspect ratio. For example, nanoparticles having that spherical-, oval-, cubic-, prism-, helical-, or pillar-shaped morphologies.

2.3 Classification Based on *Crystalline Forms*

The basic difference between crystalline nanostructures and amorphous nanoparticles lies in the way how the atoms are arranged around each other in the nanocrystalline systems besides their size. Thus on the basis of crystallinity, solids generally appear in three forms (Fig. 3)

(i) Amorphous—It does not have long range order, e.g., as glasses. They have some short range order at the atomic length scale depending on the nature of chemical bonding.
(ii) Polycrystalline—They have multiple domains of varying size and orientation. Their orientation may not have any preferred directions. Therefore, they may say to have random texture. These crystallite domains are sometime also referred to as grains.
(iii) Crystalline—A single extended domain have high long range order, properly arranged domains.

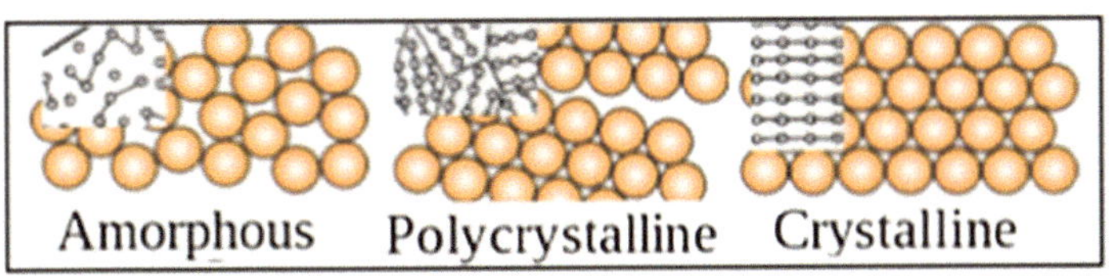

Fig. 3 Different types of crystalline forms

2.4 Classification Based on *Chemical Nature*

We know that all the materials are basically formed with the chemicals which may be inorganic, organic, or mixture of these two. Classification of nanomaterials based on the chemical nature is pictorially depicted in Fig. 4.

2.5 Classification Based on *Chemical Composition*

Chemical composition can be referred to the type of arrangement of atoms according to the ratio of the size of atoms in molecules of a particular material. Upon the addition and subtraction of chemical reagents, chemical composition of the particular substance changes. The properties of a substance are established according to its chemical composition.

According to Gleiter [3] the one-, two-, or three-dimensional nanostructures may be further classified on the basis of their chemical composition too, into four families (Fig. 5).

(i) First family—In the most simple case in which all the constituents are distributed homogenously through out the region having same chemical composition, e.g., semicrystalline polymers or multilayers of thin film crystallites which are separated by an amorphous layer, etc. [4]

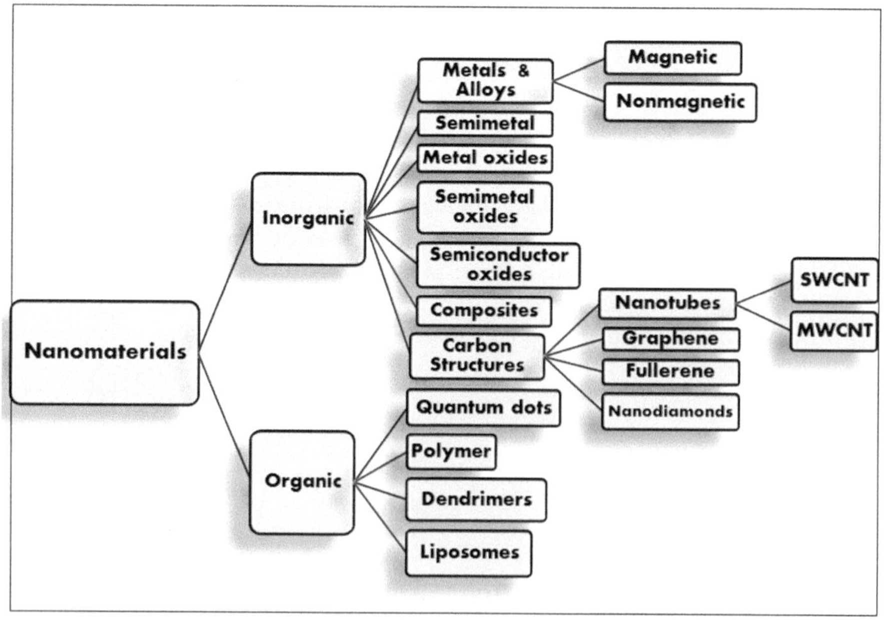

Fig. 4 Classification of nanomaterials based on chemical nature

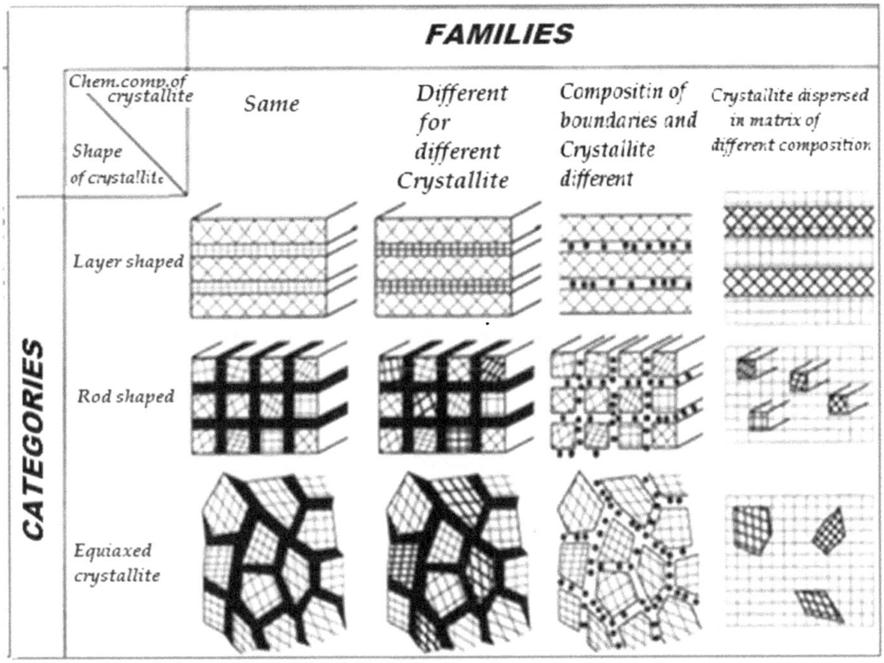

Fig. 5 Classification of nanomaterials based on chemical composition [3]

(ii) Second family—Those substances in which materials have different chemical composition of constituents, e.g., quantum well structures.
(iii) Third family—Those materials in which they a different chemical composition or having materials of mixed component system, e.g., alumina ceramic with Ga in its interface [5].
(iv) The fourth family—All those nanomaterials which are formed by nanometer-sized particles (layers, rods or equiaxed crystallites) dispersed in such a matrix having different chemical composition, e.g., hardened alloys precipitation.

2.6 Classification Based on *Magnetic Behavior*

All the materials do not respond similarly with the externally applied magnetic field. Thus, on the basis of their response to any externally applied magnetic field, nanoparticles may be classified into five categories, namely, diamagnetic, paramagnetic, ferromagnetic, antiferromagnetic, and ferrimagnetic [6]. Orientations of the magnetic moments in a material place them in the above categories accordingly. At nanolevel also, the particles exhibits magnetic properties only in the presence of an external magnet and changes to a nonmagnetic state again when the external magnetic field is removed. At this state, many factors play their role in determining the magnetic properties of nanomaterial, such as the, particle size, shape, morphology, their chemical composition and interaction with the surrounding matrix and the neighboring particles. The magnetic behavior of any material can be controlled by altering any of these factors to a certain limiting extent. For example, spontaneous magnetisation occurs below T_C when the particle size is ≥ 1 μm. For single-domain magnetic particles with particle size equal to 1–30 nm, at $T_b < T_C$, the orientation of the magnetic moment of a particle remains same in space, but the particle ensemble shows a magnetic hysteresis. At higher temperature above to T_b, the particle attains superparamagnetic state. Usual paramagnetic properties are retained by a single atom (ion) with particle size equal to $\sim$0.2 nm. At the range of the quantum dots, high field irreversibility type of new phenomena, superparamagnetism or extra anisotropy contributions, are observed by the magnetic nanoparticles. The magnetic behavior of individual nanoparticles develops such phenomena due to its small and finite-size and surface effects [7]. In a superparamagnetic material, spins are substituted by small ferromagnetic domains. Generally, when the particle comes in a state of uniform magnetization at any field, it should have only a single magnetic domain. As a result of magnetic anisotropy, such type of superparamagnetism arises. It means that the spins are aligned along a preferred crystallographic direction at this state [8]. In the presence of an external magnetic field, the aligned domains generate a strong attractive interaction. Once the external magnetic field is removed, thermal agitation cancels residual magnetization.

Despite of the great importance of the morphology of nanoparticles, in magnetic devices for storing or processing information, where well-defined magnetization axes and switching fields are required, and the alignment of domains play its important role.

2.7 *Classification Based on Functionalization*

With the advancement in the nanomaterial-based applications, the development of functionalized nanoparticles gained specific recognition especially in the field of biomedical and life sciences. Clinical results have shown that functionalized nanoparticles with different functional groups or can say multifunctional nanoparticles yielded enhanced efficacy with minimal side effects [9]. For stabilization purposes and to avoid agglomeration, nanoparticles surface modification/coating is done by some capping or functionalizing agents. Thus, the on the basis of coating/functionalization, nanoparticles can also be classified as

(i) Bare nanoparticles—without any coating
(ii) Capped Nanoparticles—to avoid agglomeration

(iii) Functionalized nanoparticles—to perform certain specific task. Functionalization may be of two types

(a) Post Functionalization—When the functionalization of nanoparticle is done by capping or grafting organic groups onto the active surface of nanomaterials after the synthesis is known as post functionalization [10].

(b) *In situ* Functionalization—When during the synthetic process only functionalization and modification of nanomaterials is carried by organic compounds, then it is said to be *in situ* functionalization [11, 12]

2.8 Classification Based on Applications

Jack et al. have well said in their book "The Next Big Thing Is Really Small" [13] that how the future of our business will be changed by nanotechnology, i.e., nanotechnology has a wide horizon of its applicatory attributes. Perhaps not a single sphere of application is left without nanotechnology. Its broad applicatory scenario may be broadly classified as mentioned in Fig. 6.

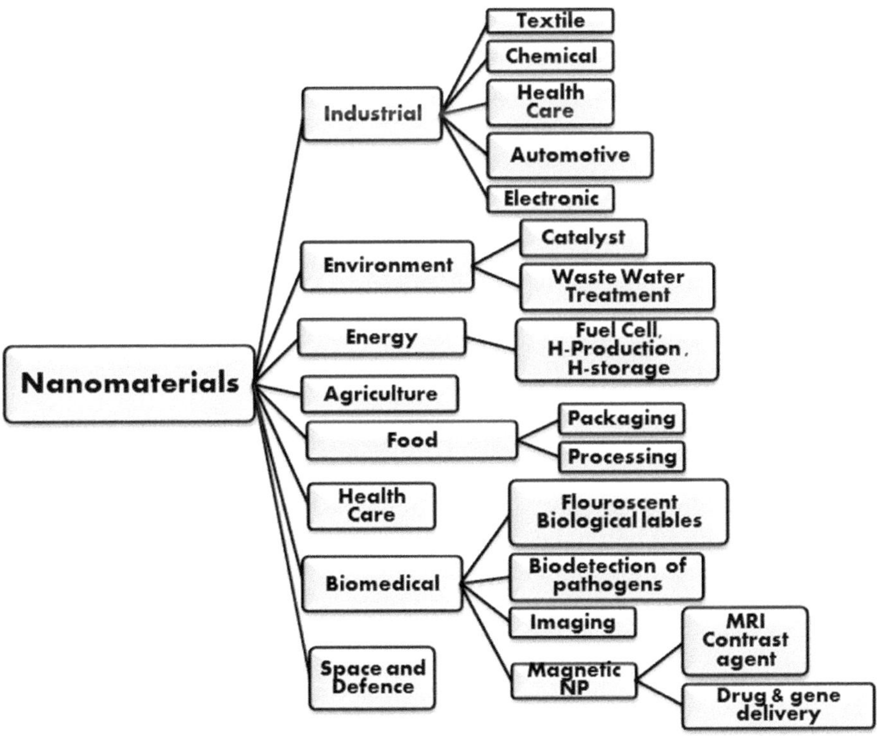

Fig. 6 Classification of nanomaterials based on applications

3 Reason for the Peculiarity in the Properties

As the sizes of a substance begins to move towards the nanoscale, its properties become unique and dramatically different from the same substances in the bulk form. For doing nanotechnology, scientists work at the nanoscale level to enable them to utilize this unique different properties, viz., physical, chemical, mechanical, and optical properties of materials that naturally occur at this scale. Properties such as melting point, fluorescence, electrical conductivity, magnetic permeability, chemical reactivity, etc., changes as a function of the size of the particle. The reason for the dramatic change in the properties of nanomaterials can be attributed to the following two facts.

3.1 Discretness of Energy Level

When we go from the macroscopic to the microscopic to nanoscopic to atomic dimension, we move from classical mechanics to quantum mechanics. Quantum mechanics gives a complete version to the world of nanometric scale, which is absolutely different from the classical Newtonian mechanics, which governs operations of everyday objects involving properties that are size-dependent. But quantum mechanics governs the interactions of very small things that range from micro-molecules to nano to atomic and subatomic particles. Nobel laureate Max Planck (won Prize for Physics in 1918), show in his famous quantum mechanics, that the particle of energy that is exchanged between matter and radiation is called quanta, a discontinuous quantity. E. Schrödinger and P. Dirac (Nobel Prize winners for Physics in 1933) have developed wave mechanical concept by incorporating physicist Louis de Broglie's matter wave concept and quantum mechanics which has driven a "revolution" in this field. Wolfgang Pauli's exclusion principle designated the states of particles, which has given highest degree of accuracy in the atomic world though this world resides on probabilities and uncertainties. A molecule is formed by the overlapping of atomic orbitals (the space where the probability of finding electron is highest), i.e., representation of electrons at discrete energy levels, can only be attained as a simulation.

The electron's energy levels in quatum dots are discrete and quantified as in an atom or in a molecule and not as in the case of an ordinary semi-conductor where electrons are spread out in energy bands (Fig. 7). The distribution of energy on these levels occurs according to the particle size of the crystal.

Thus, in nanocrystals, the electronic energy levels are discrete having finite density of states, because of the confinement of the electronic wave function to the physical dimensions of the particles and not continuous band as in the bulk. This phenomenon is known as Quantum confinement. Due to the quantum confinement, there occurs change in electrical properties of the material. Some materials that show conductivity in the bulk form may become semiconductors or poor

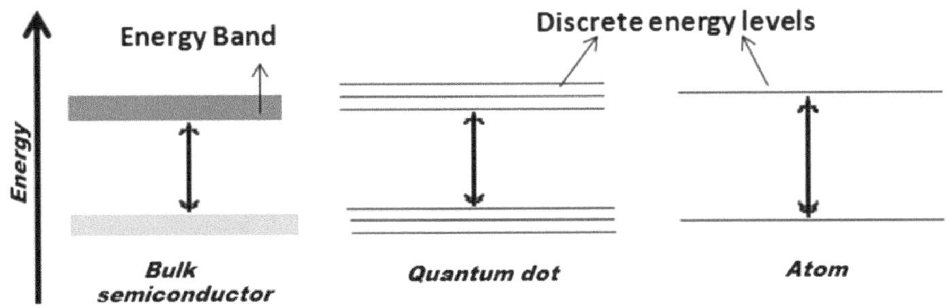

Fig. 7 Representation of the comparison of band gap energy between bulk, nano, and atom

conductors at the stage of nanometer level. Some materials that were semiconductors may become conductors or superconductors. However, the atoms at the surface experience different surrounding than those present at the center of the particle because of the presence of unsatisfied free electrons or dangling bonds. The presence of dangling bonds onto the surface of a material can change its conducting properties.

Optical properties are also size-dependent at the nanoscale level. As said above, confinement leads to a transition from continuous to discrete energy levels (Fig. 7). Due to the discreteness of energy levels, the excitonic band gap increases which restrict the movement of electrons. It cannot move about as freely. Therefore, the quantum confinement of the electrons changes the optical properties of a substance, i.e., it reacts differently to light. As the band gap increases, those substances which are usually opaque in bulk becomes transparent to light at nanolevel. For example, at the macro scale gold appears yellow in bulk but the color changes to red at nanosized scale. Similarly, large-sized zinc oxide particles are used for sun screen because it scatters visible light and appear white but at nanolevel zinc oxide particles do not scatter visible light and becomes transparent to sunlight. With the decreasing particle size, quantum dots changes the appearance of a substance by creating different phosphor and fluorescent colors.

It was found that magnetic nanoparticle can often be described as a single magnetic domain with the uniaxial anisotropy (which comes from the shape of the particle and effect of surface spins) even in materials with other type of anisotropy (such as cubic). The orientation of its magnetic moment points may be either up or down in the absence of magnetic field. During the synthesis of a crystal, there may develop several types of defects on the surface as well as in the core, such as atomic vacancies, changes in the atomic coordination, dangling bonds or lattice disorder. Due to these defects some electronic spins remain uncompensated which leads to the surface magnetization (ferromagnetism or antiferromagnetism) and again, it depends on the size of the particle and on the extent degree of disorder at the surface [14, 15]. Small particles can have very high magnetic susceptibility with permanent magnetic dipole. Small clusters consisting of a single ferromagnetic domain that follow the applied field freely are placed in the category of super paramagnetism. The magnetic susceptibility of superparamagnetic particles is in orders of magnitude larger than bulk paramagnetic materials.

The deciding factor of the properties of materials such as conductivity, magnetism, and reactivity is the electrons possessed by the materials. Electromagnetic force is unaffected by mass, as it is a function of charge and distance. This becomes even more stronger at the nanosize level. Since electromagnetic forces operate between protons and electrons, therefore as the distance between these charged particles decreases, they experience stronger electromagnetic attraction between them and vice versa. The forces that act on atoms can be attractive or be repulsive. It means that it is these columbic attractive forces that helps an atom in giving its shape and size which in-turn reflects the ability of an atom to form chemical bonds that depends on the attractive forces acting on the outermost valence electronic energy level of an atom. An atom that has incomplete outer energy level attains stability by the rearrangement of valence shell electrons by transfer or sharing of electrons with any other atom. During these rearrangements, potential energy for a chemical bond develops which is responsible for the interactions between atoms and molecules. Therefore, it is very important for the development of nanotechnology. Covalent and ionic bonds are much stronger than hydrogen bond and van der Waals forces. At the nanoscale, van der Waals forces become very prominent. Due to which materials become sticky. That is why nanofibre can be very effective in attracting and trapping small particles. This makes nanofibres excellent materials for use in filtration.

3.1.1 Increased Surface-to-Volume Ratio

We know that as the size of particle decreases, the surface area and therefore, surface-to-volume ratio also increases. This increased surface area of nanoparticles is responsible for the development of peculiar properties of a substance at the nanoscale level (Fig. 8). Due to large surface area there occurs change in the reaction time of a substance. As the particle size decreases, the percentage of atoms on their surface increases [16]. This accounts for the increase in the surface-to-volume ratio. The higher surface-to-volume ratio increase the rate of reaction due to the drastic increase in the amount of exposed surface area at the nanoscale level causing the increase in the reaction rate for a chemical reaction. Nanomaterials are inherently unstable due to high surface energy. Atoms exist at the surface or interface are different from the same atoms exist in the interior of a material. At surface due to free dangling bonds nanomaterial acquires high surface energy which make them highly unstable causing agglomeration.

The properties such as melting point, rate of reaction, capillarity and adhesion, etc., are controlled by their surface area. For example, Gold at the macro scale, has melting point of 1064 °C [18]. Its melting point radically drops about 100 °C as its particle size reduces from 100 to 10 nm diameter. On the further size reduction to

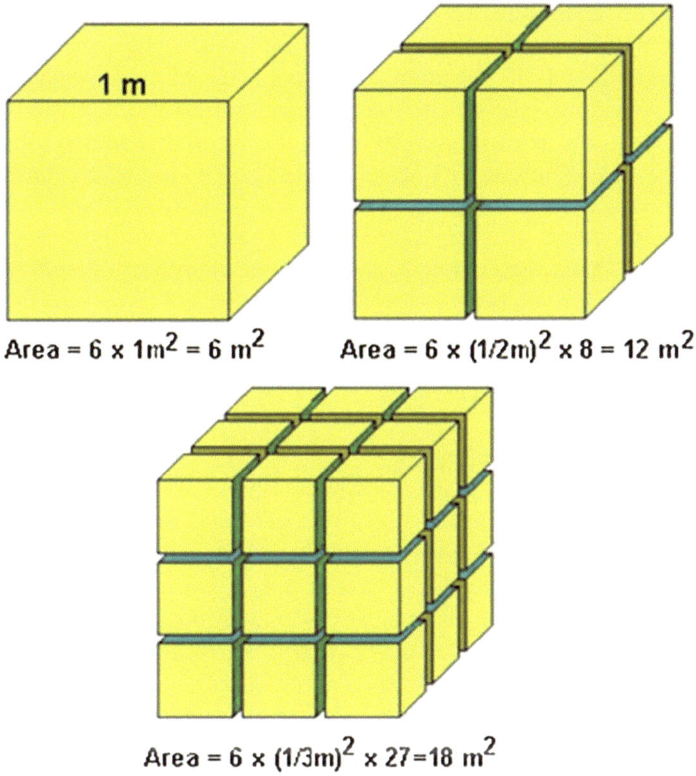

Fig. 8 Representation of increasing area as the particle size decreases [17]

about 2 nm the melting temperature reduces to about half of the melting temperature at the macro scales level. At this level gold will no longer be able to conduct electricity.

4 Magnetization Dynamics of Nanomaterials

In order to study magnetic properties and magnetization dynamics, electron magnetic resonance (EMR) or electron spin resonance (ESR) spectroscopy is preferred. Earlier a conventional ferromagnetic approach, where dynamics of the magnetization was defined as ferromagnetic resonance (FMR) with the help of Landau-Lifshits equation, considering strong thermal fluctuations have been used to study EMR of nanoparticles [19, 20]. The behavior of nanomagnetic objects lie at the boundary line between quantum dynamics and classical thermodynamics. Considering magnetic nanoparticle as a giant spin, Noginova et al. [21] have assumed that the EMR signals are resulted as a collective contributions of quantum transitions that occurred between energy levels developed due to the projections of the giant spin onto the direction of the magnetic field. Such nanoparticles gives peculiar features in EMR signals that are common in quantum systems, as the variations in narrow spectral component and multiple-quantum transitions with temperature. As the particle size increases, the relative intensity of such EMR signals decreases significantly. This suggests a route of a gradual transition from a purely quantum system applied to sub atomic state to a classical behavior of comparatively large systems, where such effects diminishes. Noginova and co-workers [22] have also explored the effects of the particle size and magnetization dynamics, with EMR studies of magnetic nanoparticles to determine their potential to use them as building blocks for tunable microwave metamaterials.

5 Biological Importance

Molecular or cellular parts of biological systems lie in the submicron size domain, As the size of the nanoparticles lie within these dimensions, they viable to have number of biomedical applications, such as they can be exploited as a very small probes to spy within the cellular machinery without disturbing or interfering their working conditions. They can be easily capped with biological molecules, such as DNA, RNA, proteins or carbohydrates, to enable them to interact with or bind to a biological moieties, i.e., by "tagging" or "addressing" they can provide a controllable means of delivering drug of in diagnostic applications. By changing gradient of external magnetic field, magnetic nanoparticles can be taken and assembled to the desired target. Since human tissues have intrinsic penetrability to the magnetic fields, with such type of external control mechanism operating from an optimum distance, opens up many applications which involves transportation and immobilization of these magnetic nanoparticles, or magnetically tagged biological species. In such cases, the external magnetic field assists the injected magnetic nanoparticles along with the tagged drug to the desired site within the body acting as a site-specific drug delivery busses. Magnetic nanoparticles such as gadolium may be used as contrast agents in magnetic resonance imaging (MRI). On subjecting to an external magnetic AC field, the temperature around the effected sites can be raised up to 40 °C as a hyperthermic cancer treatment destroying the cancerous tissues and tumors [23].

Since magnetic nanocomposites have shown to have fluorescent, radio-opaque and paramagnetic properties, they may be utilized in confining and concentrating the target analytes in a very small minute volume for in situ optical detections. Therefore, magnetic nano particles functionalized or tagged with biomolecules such as DNA, proteins, peptides or antibodies significantly improves the selectivity and sensitivity of the system serving as bio- and chemo-sensors [24, 25]. Corr et al. have utilized electrostatic interactions to develop new fluorescent-magnetic nanocomposites. They have prepared fluorescent magnetite-porphyrin nanocomposites by interacting core nanoparticle, the spacer group and the fluorophore. Intracellular fragmentation of the above said nanocomposite have shown their utility as subcellular imaging contrast agents and in targeted drug delivery systems [26].

Magnetic iron oxide nanoparticles have thus shown a very wide spectrum of applications in various clinical, diagnostics and therapeutic fields such as, cancer, cardiovascular, neurological disorders, hyperthermia, magnetic separation of cells, proteins, DNA/RNA, and in other biomolecular probes. Although they have been found to be very useful in various biomedical and in vitro applications but because of the agglomeration and necessity of fast detection via immune cells in physiological medium, their clinical use are mired [27–31]. In order to use nanoparticles in biological systems, its water-dispersibility in bio-systems is very important and it should be bio-compatible. The colloidal stability of nanoparticle suspension should also be maintained in the bio-physiological conditions, with a wide range of pH [9].

6 Conclusion

Physicist Richard Feynman ignited the torch of revolution called nanotechnology and it has illuminated the minds all over the world. Now it has developed number of tentacles. Nanomaterials accordingly classified in number of ways and its applications in many fields are discussed. At the nanoscale level, quantum mechanics plays very important role in determining the properties and characteristics of a material which leads to change in its optical, electrical and magnetic behaviors. The properties of the material become more or less surface dependent. Due to the increase in the surface-to-volume ratio at the nanoscale many surface-dependent properties show changes in many folds. When nanomaterials are engineered in the right way, they not only become electrifying in terms of applications, but also become very significant in current and fast growing field of nanotechnology.

Acknowledgments We gratefully thank "Nanotechnology application center" for facilitating us to work and for literature survey. We acknowledge the valuable suggestions and encouragement given by Dr. Lalit Eusebius, Head, Department of Chemistry, Ewing Christian College, Allahabad.

References

1. Taniguchi, N.: On the basic concept of 'nano-technology'. In: Proceedings of International Conference Production Engineering, Tokyo, Part II, Japan Society of Precision Engineering (1974)
2. Siegel, R.W.: Nanophase materials, in encyclopedia of applied physics. In: Trigg, G.L. (ed.) vol. 11, pp. 1–27. Weinheim, VCH (1994)
3. Gleiter, H.: Acta Mater. **48**, 1–29 (2000)
4. Dutta, J., Reaney, I.M., Cabarrocas, P.R.I., Hofmann, H.: Nanostruct. Mater. **6**, 843 (1995)
5. Konrad, H., Weissmüller, J., Hempelmann, J., Birringer, R., Karmonik, C., Gleiter, H.: Phys. Rev. B **58**, 2142 (1998)

6. Gignoux, D.: Phenomenology of Magnetism at the Macroscopic Scale. Springer, New York, NY, USA (2005)
7. Grancharov, S.G., Zeng, H., Sun, S.H., Wang, S.X., et al.: Bio-functionalization of monodisperse magnetic nanoparticles and their use as biomolecular labels in a magnetic tunnel junction based sensor. J. Phys. Chem. B. **109**(26), 13030–13035 (2005). doi:10.1021/jp051098c
8. Kumar, R.V., Koltypin, Y., Cohen, Y.S., Cohen, Y., Aurbach, D., Palchik, O., Felner, I., Gedanken, A.: J. Mater. Chem. **10**, 1125 (2000)
9. Sanjay, S.S., Pandey, A.C.: Functionalization of smart nanomaterials. In: Tiwari, A., Kobayashi, H. (eds.) Responsive Materials and Methods, pp. 201–236. Scrivener Publishing LLC (2014)
10. Kickelbick, G., Schubert, U.: Synthesis, functionalization and surface treatment of nanoparticles. In: Baraton, M.I. (ed.) American Scientific Publishers, Stevenson Ranch, CA (2003)
11. Sanchez, C., Illia, G.S., Ribot, F., Lalot, T., Mayer, C.R., Cabuil, V.: Chem. Mater. **13**, 3061 (2001)
12. Yin, S., Aita, Y., Komatsu, M., Wang, J., Tang, Q., Sato, S.: J. Mater. Chem. **15**, 74 (2005)
13. Uldrich, J., Newberry, D.: Crown Publishing Group(US), ISBN 9781400049172 (2003)
14. Ho, C.-H., Lai, C.-H.: Size-dependent magnetic properties of PtMn nanoparticles. IEEE Trans. Magn. **42**, 3069–3071 (2006)
15. Dobrynin, A.N., Ievlev, D.N., Temst, K., Lievens, P., Margueritat, J., Gonzalo, J., Afonso, C. N., Zhou, S.Q., Vantomme, A., Piscopiello, E., van Tendeloo, G.: Critical size for exchange bias in ferromagnetic-antiferromagnetic particles. Appl. Phys. Lett. **87**, 012501 (2005)
16. Batlle, X., Labarta, A.: Finite-size effects in fine particles: magnetic and transport properties. J. Phys. D **35**, R15–R42 (2002)
17. http://www.uwgb.edu/dutchs/GRAPHIC0/GEOMORPH/SurfaceVol0.gif
18. Castro, T., Reifenberger, R., Choi, E., Andres, R.P.: Size-dependent melting temperature of individual nanometer-sized metallic clusters. Phys. Rev. B **13**, 8548–8556 (1990)
19. Raikher, Y.L., Stepanov V.I.: Phys. Rev. B **50**, 6250 (1994)
20. De Biasi, E., Ramos, C.A., Zysler, R.D.: J. Magn. Magn. Mater. **262**, 235 (2003)
21. Noginova, N., Chen, F., Weaver, T., Giannelis, E.P., Bourlinos, A.B., Atsarkin,V.A.: Magnetic resonance in nanoparticles: between ferro- and paramagnetism 2007. J. Phys. Condens. Matter. **19**, 246208 (2007)
22. Noginova, N., Quincy, L.W., Panagiotis, D., Giannelis, E.P.: Magnetic nanoparticles for tunable microwave metamaterials. In: SPIE NanoScience + Engineering, pp. 845531–845531. International Society for Optics and Photonics, 2012
23. Kumar, R.V., Koltypin, Y., Cohen, Y.S., Cohen, Y., Aurbach, D., Palchik, O., Felner, I., Gedanken, A.: J. Mater. Chem. **10**, 1125 (2000)
24. Katz, E., Willner, I.: Integrated nanoparticle-biomolecule hybrid systems: synthesis, properties, and applications. Angew. Chem. Int. Ed. **43**, 6042–6108 (2004)
25. Yang, H.S., Santra, S., Walter, G.A., Holloway, P.H.: GdIII-functionalized fluorescent quantum dots as multimodal imaging probes. Adv. Mater. **18**, 2890–2894 (2006)
26. Corr, S.A., O'Byrne, A., Gun'ko, Y.K., Ghosh, S., Brougham, D.F., Mitchell, S., Volkov, Y., Prina-Mello, A.: Magnetic-fluorescent nanocomposites for biomedical multitasking. Chem. Commun. **43**, 4474–4476 (2006)
27. Sun, C., Lee, J.S., Zhang, M.: Magnetic nanoparticles in MR imaging and drug delivery. Adv. Drug Deliv. Rev. **60**(11), 1252–1265 (2008)
28. Saltan, N., Kutlu, H.M., Hür, D., Işcan, A., Say, R.: Interaction of cancer cells with magnetic nanoparticles modified by methacrylamido-folic acid. Int. J. Nanomed. **6**, 477–484 (2011)

29. Hanini, A., Schmitt, A., Kacem, K., Chau, F., Ammar, S., Gavard, J.: Evaluation of iron oxide nanoparticle biocompatibility. Int. J. Nanomed. **6**, 787–794 (2011)
30. Oh, J.K.: Iron oxide-based superparamagnetic polymeric nanomaterials: design, preparation, and biomedical application. Prog. Polym. Sci. **36**(1), 168–189 (2011)
31. Xie, Y., Zeng, P., Siegel, R.A., Wiedmann, T.S., Hammer, B.E., Longest, P.W.: Magnetic deposition of aerosols composed of aggregated superparamagnetic nanoparticles. Pharm. Res. **27**(5), 855–865 (2010)

An Overview on Advances in the Nanocarriers Drug Delivery Systems

Anjana Pandey

Abstract The unceasing efforts and improvement of drug delivery systems (DDSs) have been broadly researched to maximize therapeutic efficacy while curtailing undesirable side effects. Nanoparticle technology was recently shown to hold great promise for drug delivery applications in nanomedicine due to its favorable properties, such as better encapsulation, bioavailability, control release, and lower toxic effects. Regardless of the great progress in nanomedicine, there remain many limitations prior to widely being accepted for medical application. To overcome these limitations, advanced nanoparticles for drug delivery have been developed to enable the spatially and temporally controlled release of drugs in response to specific stimuli at disease sites. An ideal drug delivery system should be able to localize a drug specifically and directly to its target. This is particularly important when drugs made by traditional manufacturing methods are hydrophobic and their solvents are toxic. Nanotechnology promises to improve drug delivery system design and targeting. Nanostructured drugs or delivery carriers allow the continuous and controlled release of therapeutic drugs to maintain drug levels to a desired extent. The size of nanoparticles ranges from 10 to 200 nm, about the size of a protein. Because of their small size, nanoparticles can readily interact with biomolecules on the cell surface or inside cell allowing these nanoparticles to penetrate tissues in depth with a high level of specificity. This chapter discusses an overview of nanoparticulate systems that can be used as a potential drug delivery carriers and focuses on the potential applications of nanoparticles in various biomedical fields for improving human health care.

Keywords Drug delivery · Nanoparticles · Biomolecules

A. Pandey (✉)
Department of Biotechnology, Motilal Nehru National Institute of Technology (MNNIT), Allahabad, UP 211004, India
e-mail: anjanap@mnnit.ac.in

A.K. Shukla (ed.), *EMR/ESR/EPR Spectroscopy for Characterization of Nanomaterials*, Advanced Structured Materials 62,
DOI 10.1007/978-81-322-3655-9_3

1 Introduction

An upcoming area of nanotechnology is "nanomedicine," which employs molecular understanding of a drug and its binding surface to address medical problems. Nanotechnology deals with the design, synthesis, characterization, and application of materials and devices at the nanoscale (1–100 nm) [5, 15] which show novel properties from their bulk materials [16]. The scope of nanomedicine ranges from in vivo imaging and diagnosis to therapeutics such as drug delivery and gene therapy. Nanotechnology has immense potential in diversified areas of health-related areas, for example, nanoparticle (NP)-based assay devices have been demonstrated to be able to detect proteins in the attomolar concentration range (protein biomarkers, toxins etc.), a magnitude of six orders lower than concentrations detected by ELISA [14]. Various materials (e.g., polymer, inorganic precursors, silica, noble metal, and carbon) could be used for the fabrication of nanoparticles which have the ability to load a wide range of bioactive components, including small molecular chemotherapeutics, proteins, nucleic acid, and diagnostics, for biomedical application [12]. Besides that functional groups (such as –OH, –COOH, $-NH_2$, etc.) of the nanoparticles make them suitable for conjugation of drugs, genes, antibody, and aptamer allowing them for targeted imaging and treatment of various diseases [8]. Nanoparticles are taken up by cells more efficiently by diffusion or through receptor mediated than larger micron-sized molecules and therefore, could be used as effective transport and delivery systems. For, therapeutic applications, drugs can either be integrated in the matrix of the particle or attached to the particle surface. A drug targeting system should be able to control the fate of a drug entering the biological environment. Nanosystems with different compositions and biological properties have been extensively investigated for drug and gene delivery applications [9, 19, 39, 48, 50]. An effective approach for achieving efficient drug delivery would be to rationally develop nanosystems based on the understanding of their interactions with the biological environment, target cell population, target cell surface receptors [17, 27, 37], changes in cell receptors that occur with progression of disease, mechanism, and site of drug action, drug retention, multiple drug administration, molecular mechanisms, and pathobiology of the disease under consideration. It is also important to understand the barriers to effective drug delivery such as stability of therapeutic agents in the living cell environment. Reduced drug efficacy could be due to instability of drug inside the cell, unavailability due to targeting multiple locations or chemical properties of the delivery molecules, alterations in genetic makeup of cell surface receptors, over-expression of efflux pumps, changes in signaling pathways with the progression of disease, or drug degradation. For example, excessive DNA methylation with the progression of cancer [18] causes failure of several anti-neoplastic agents like doxorubicin and cisplatin. A better understanding of the mechanism of uptake, intracellular trafficking, retention, and protection from degradation inside a cell are required for enhancing efficacy of the encapsulated therapeutic agent.

This improvement will allow early detection of many diseases such as cancer, genetic disorders, cardiovascular diseases, infectious diseases, etc., and may save millions of lives through the prevention and early treatment of these diseases. Synthesis of nanodrug and/or gene delivery system provides superior potential for many applications, including antitumor therapy by targeted delivery of therapeutic agents to tumors. Treatment of cancer represents an enormous biomedical challenge for drug delivery. A nanoparticle-mediated drug delivery system can significantly eliminate drug or drug carrier side effects. Targeted drug delivery creates better therapeutic effect than that of conventional drug dosing method with minimized side effects because of reduced dosage required at target site. An ideal drug delivery system should be able to localize a drug specifically and directly to its target. This is principally important when drugs made by conventional manufacturing methods that are hydrophobic and their solvents are toxic. Nanotechnology promises to improve drug delivery system design and targeting because of biodegradability, target site uptake, increased surface area, and stimuli responsive drug release. Nanostructured drugs or delivery carriers allow the continuous and controlled release of therapeutic drugs to maintain drug levels within a desired level in response to various stimuli like pH, light, magnetic field, redox potential, ultrasound waves, temperature, ionic strength, enzyme, and substrate concentration. Other advantages of nanostructured drugs include localizing and specifically targeting the drugs to their intended tissues and cells, thereby decreasing drug doses and improving patient compliance because of minimized side effects of drug. In this chapter, we will describe a variety of possible nanodrug carriers.

2 Types of Nanoparticles Used in Drug Delivery

Nanoparticles applied as drug delivery vehicles are generally less than 100 nm in at least one dimension. These nanocarriers may consist of different biodegradable materials such as natural or synthetic polymers, lipids, metal oxides, or metals.

2.1 Polymer-Based Drug Carriers

Depending on the method of preparation, the drug is either physically entrapped in or covalently bound to the polymer matrix [44]. The resulting compounds may have the structure of capsules (polymeric nanoparticles), amphiphilic core/shell (polymeric micelles), or dendrimers (Fig. 1a, b, d). Polymers used as drug conjugates can be divided into two groups of natural and synthetic polymers. Polymers such as albumin, chitosan, and heparin occur naturally and are nontoxic polysaccharide polymers [32]. These are also biodegradable and biocompatible and hence been the materials of choice for the delivery of oligonucleotides, DNA, protein as well as the drugs. Biodegradable polymeric nanoparticles, mainly consisting of polylactic acid

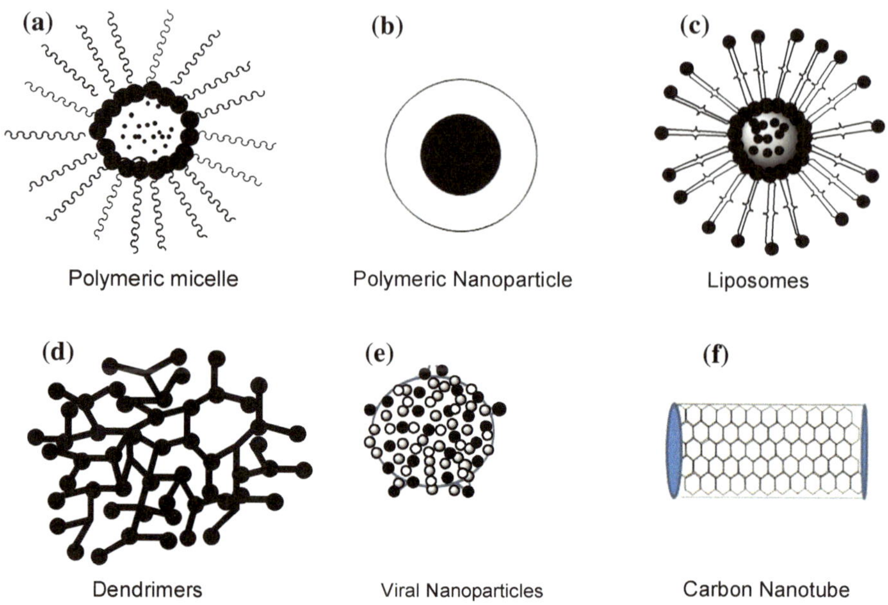

Fig. 1 Types of nanocarriers for drug delivery. **a** Polymeric micelle: The hydrophobic core region encapsulates the hydrophobic drug while hydrophilic shell renders the micelle water soluble. **b** Polymeric nanoparticles: Drugs are conjugated to or encapsulated in polymers. **c** Liposomes: Lipid bilayer spherical structure with central aqueous region for encapsulation. **d** Dendrimers: Highly branched synthetic polymeric macromolecule of nanometer dimension. **e** Viral nanoparticles: protein cages which are multivalent self-assembling structure. **f** Carbon nanotube: carbon cylinders comprised of benzene rings

(PLA), polyglycolic acid (PGA), polylactic-glycolic acid (PLA-PGA), and poly (methyl methacrylate) (PMMA) have been used for both gene and drug delivery [58].

2.2 *Liposomes*

Liposomes are spherical self-assembling lipid particles in which an outer lipid bilayer encloses a core aqueous space (Fig. 1c). The lipid layers of liposome in fact protect and entrap the drug until the liposome binds to the outer membrane of target cells. Nanocarriers made with liposomes have a long circulation time, reduced systemic toxicity, increased uptake into tumors, and steady release of their payloads [41, 59]. Liposomes are one of the first nanoparticle-based drug delivery platforms to be applied in medicine [6]. The best example of a clinical liposome formulation is liposomal doxorubicin (Doxil™). The encapsulation of doxorubicin in liposomes shielded by polyethylene glycol (PEG) prolongs systemic drug circulation, improves the therapeutic efficacy, and minimizes off target effects relative to the free drug [3, 24].

2.3 *Dendrimers*

Dendrimers are synthetic polymeric macromolecule of nanometer dimension (Fig. 1d). They are made up of different types of polymers such as poly(L-glutamic acid) (PGA), polyamidoamine (PAMAM), poly(ethylene glycol) (PEG), and polyethylenimine (PEI) that emerge radially from the central core using either convergent or divergent step growth polymerization [1]. The dendrimer is hydrophilic and so can preferably be used as a coating agent for drug delivery. The process of preparation of dendrimers is quite easy and its terminal can be modified for targeted drug delivery and selective imaging of different tissues such as tumors [42, 43].

2.4 *Viral Nanoparticles*

Some viruses (canine parvovirus, cowpea chlorotic mottle virus, cowpea mosaic virus, and bacteriophages) have been developed for biomedical and nanotechnology applications like tissue targeting and drug delivery (Fig. 1e) [31]. A number of functional targeting molecules and peptides can be displayed on the virus capsid using chemical or genetic engineering. Several antibodies, ligands (transferrin and folic acid) have been conjugated to viruses for specific tumor targeting *in vivo* [45].

2.5 *Superparamagnetic Nanoparticles*

In recent years magnetic inorganic nanomaterials have gathered considerable attention because of their potential for biomedical applications such as targeted drug delivery [28, 53] including cancer diagnosis and treatments [25]. The magnetic properties of nanometer sized iron oxide powders such as α-Fe_2O_3, γ-Fe_2O_3, and Fe_3O_4 have been extensively studied [20, 61] and compared with their bulk counterparts. Because of their high tissue permeability and colloidal stability, applications of these magnetic nanoparticles as drug carriers and for controlled drug release within the target cells, are enormous. Furthermore, these oxide materials exhibit better chemical stability and biocompatibility in comparison to many metallic magnetic materials [25]. In addition, superparamagnetic nanoparticles of iron oxides do not retain magnetization before and after exposure to an external magnetic field, which reduces the possibility of aggregation of particles in a magnetic dispersion.

2.6 Gold Nanoparticles

Gold nanoparticles have emerged as nontoxic drug carriers for selective delivery of drugs into their targets. PEGylation is a common method to increase stability and robustness of these nanoparticles. The release of the encapsulated therapeutic agent could be triggered by internal (e.g., glutathione or pH) or external (e.g., light) stimuli. Gold nanoparticles exploit their unique chemical and physical properties for transporting and releasing the pharmaceuticals [57]. First, the gold core is essentially inert and nontoxic. The second advantage is their easy synthesis procedure with monodisperse nanoparticles formation. Further, these nanoparticles are versatile because of their functionalization through a thiol linkage.

2.7 Carbon Nanotubes

Carbon nanotubes are carbon cylinders composed of six membered rings (Fig. 1f) that can be used as sensors, diagnostic devices, and carriers to deliver drugs. Carbon nanotubes are completely insoluble in all solvents, generating some health concerns and toxicity problems. Introduction of functional groups to carbon nanotubes can make them water soluble and functionalized so that the active molecules like peptides, proteins, nucleic acids, and drugs can be linked to it [7]. Methotrexate, an

Table 1 Types of nanocarriers for drug delivery

Particle type therapeutic application	Materials	Size (nm)	Toxicity
Magnetic nanoparticles diagnosis, disease treatment and drug delivery	Iron oxide	5–100	Low
Gold nanoparticles drug delivery	Chloroauric acid sodium citrate	30 nm–1.6 μm	Low
Dendrimers drug/gene delivery	Branched polymers	5–50	Polymer and cell type dependent, low
Polymeric nanoparticles drug/gene delivery	Polylactic acid, polyglycolic acid, polylactic-glycolic acid (PLGA), chitosan and poly (methyl methacrylate) (PMMA)	50–2,000	Low
Liposomes drug/gene delivery	Lipid mixtures	30–200	Low
Viral particles gene delivery	Protein cage	30–100	High
Carbon nanotube drug/gene delivery	Carbon cylinders of benzene ring	100–500	High

anticancer drug, has been linked covalently to carbon nanotube with fluorescein isothiocyante (FITC). Therefore, FITC-linked drug-carbon nanotubes were shown to be more effectively internalized through folate receptors into cells in comparison to the free drug [38] (Table 1).

3 Size and Functionalization of Nanoparticles

For effective delivery of drug to the target tissue, nanoparticles should have the ability to remain in the blood stream for a substantial time (6–10 h) without being eliminated. The injected nanoparticles are usually trapped in the circulation by the reticuloendothelial system, such as the liver and the spleen, depending on their size and surface characteristics [34].

3.1 Size

The size of nanoparticles used in a drug delivery system should be large enough to prevent their rapid leakage into blood capillaries but small enough to escape capture by fixed macrophages that are lodged in the reticuloendothelial system, such as the liver and spleen. The size of the sinusoid in the spleen and fenestra of the Kuffer cells in the liver are in the range of 150–200 nm [60] and the size of gap junction between endothelial cells of the leaky tumor vascular system lies between 100 and 600 nm [62]. As a result, the size of nanoparticles should be up to maximum of 100 nm (in exception to carbon nanotubes which are larger than 100 nm) to reach tissues by passing through these two particular vascular structures.

3.2 Surface Functionalization

The surface property of nanoparticle is an important factor which decides the life span and fate during circulation. Nanoparticles should ideally have a hydrophilic surface to escape macrophage capture [33]. This can be achieved by two ways: coating the surface of nanoparticles with a hydrophilic polymer, such as PEG and its derivatives, protects them from opsonization by repelling plasma proteins; on the other hand, nanoparticles can be synthesized from copolymers with hydrophilic and hydrophobic domains [2, 23].

For successful nanoparticle-mediated drug delivery, nanoparticles should be able to target specific cells or tissues. Therefore, it is critical to modify the surface properties of nanoparticles to achieve targeted drug delivery [29, 51]. Specific ligands such as monoclonal antibodies, folic acid and peptides can be utilized as suitable target molecule for targeted drug delivery in cancer therapy [11, 21, 22, 52].

These ligands can specifically interact with tumor-associated antigens or receptors. For instance, nanoparticles conjugated with aptamer have shown a very high specificity for drug delivery in prostate cancer therapy. The A 10 2′-fluoropyrimidine RNA aptamer recognizes the extracellular domain of the prostate-specific membrane antigen (PSMA), an antigen expressed on the surface of prostate cancer cells. These surface-functionalized nanoparticles can selectively deliver the docetazel (Dtxl) (anti-prostate cancer drug), to prostate cancer cells through the interaction between the nanoparticles-conjugated aptamers and PSMA. The drug Dtx1 was encapsulated in PLGA-PEG copolymer nanoparticles, and the carboxylic acid group present at the terminal end of PEG provides negative charge used for conjugating amine-functionalized A 10 PSMA aptamer using carbodiimide coupling chemistry [10, 13].

In order to further increase the selectivity of therapeutic nanoparticles, a two-ligand approach has been worked out. This approach has been tested in the human KB cell lines (Keratin forming tumor cell line HeLa), which overexpress both epidermal growth factor receptor (EGFR) and folate receptor (FR). Liposome-based nanoparticles bearing folic acid and a monoclonal antibody against EGFR were designed to selectively deliver doxorubicin to tumor cells [46].

Another method selectively targeted drug delivery can be achieved by the drug entrapped polymeric micelles formed by conjugates of soluble copolymers with lipids for example, the poly(ethylene glycol)-phosphatidyl ethanolamine conjugate (PEG-PE). These can spontaneously target those parts of the body area which have the compromised vasculature (tumors, infarcts) via the enhanced permeability and sustainability of the system [54]. In addition, micelles formed by lipid core and positively charged lipids are capable of escaping endosome capture and delivering drugs directly into a cellular cytoplasm [56].

Biocompatible and biodegradable polylactic-glycolic acid (PLGA) has been the most widely investigated biomaterial for making nanoparticles for controlled release and sustainable drug delivery [26, 47]. PLGA undergoes biodegradation upon implantation into the body by the citric acid cycle [40]. Furthermore, the degradation rate of PLGA can be controlled by changing block copolymer (unit polymer) composition and molecular weight which accordingly regulates the release rate of encapsulated drugs [30]. The controlled drug release from biodegradable polymer nanoparticles can be further improved by using a "nanocell" delivery system, comprised of nuclear nanoparticles (core of the nanocell) made from PLGA to which drug can be conjugated and coated with a nanoscale PEGylated phospholipid block-copolymer envelop to form a "nanocell" structure [49]. This type of "nanocell" nanoparticles is an effective drug delivery system for treatment of tumors. Furthermore, the combination of two drugs can also be delivered using "nanocells" is an exciting new advance in suppressing tumor growth. For example, trastuzumab (Herceptin) and rituximab (Rituxane) have been linked to poly(lactic acid) nanoparticles resulting in composites that exhibit sixfold enhanced nanoparticles uptake compared without targeting molecules [35, 36].

At last but not the least, another significant advantage of nanoparticle in drug delivery is the treatment of age-related neurodegenerative disorders because of unique capability of nanoparticles to transport across the blood–brain barrier (BBB) by passive diffusion or carrier-mediated transcytosis pathways [4, 55].

4 Conclusion

Nanoparticles mediated drug delivery has emerged as a promising field of biomedical applications of nanoparticles. Their unique properties like high surface area, tunable stability, functionality, and low inherent toxicity make them materials of choice to deliver drugs effectively at the target site. Development of multi-functional nanoparticles remains to be explored for medical applications along with toxicological issues before being applied to humans.

References

1. Aulenta, F., Drew, M.G., Foster, A., Hayes, W., Rannard, S., Thornthwaite, D.W., Worrall, D.R., Youngs, T.G.: Synthesis and characterization of fluorescent poly(aromatic amide) dendrimers. J. Org. Chem. **70**, 63–78 (2005)
2. Adams, M.L., Lavasanifar, A., Kwon, G.S.: Amphiphilic block copolymers for drug delivery. J. Pharm. Sci. **92**, 1343–1355 (2003)
3. Adiseshaiah, P.P., Hall, J.B., McNeil, S.E.: Nanomaterial standards for efficacy and toxicity assessment. WIREs Nanomed. Nanobiotechnol. **2**, 99–112 (2010)
4. Alyaudtin, R.N., Reichel, A., Lobenberg, R., Ramge, P., Kreuter, J., Begley, D.J.: Interaction of poly(butyl cyanoacrylate) nanoparticles with the blood-brain barrier in-vivo and in vitro. J. Drug Targeting **9**, 209–221 (2001)
5. Amoozgar, Z., Yeo, Y.: Recent advances in stealth coating of nanoparticle drug delivery systems. WIREs Nanomed. Nanobiotechnol. **4**, 219–233 (2012)
6. Bangham, A.: Liposomes: the babraham connection. Chem. Phys. Lipids. **64**, 275–285 (1993)
7. Bianco, A., Kostarelos, K., Prato, M.: Applications of carbon nanotubes in drug delivery. Curr. Opin. Chem. Biol. **9**, 674–679 (2005)
8. Biju, V.: Chemical modifications and bioconjugate reactions of nanomaterials for sensing, imaging, drug delivery and therapy. Chem. Soc. Rev. **43**, 744–764 (2014)
9. Brannon-Peppase, L., Blanchette, J.Q.: Nanoparticle and targeted systems for cancer therapy. Adv. Drug Deliv. Rev. **56**, 1649–1659 (2004)
10. Cheng, J., Teply, B.A., Sherifi, I., Sung, J., Luther, G., Gu, F.X., Levy-Nissenbaum, E., Radovic-Moreno, A.F., Langer, R., Farokhzad, O.C.: Formulation of functionalized PLGA-PEG nanoparticles for in vivo targeted drug delivery. Biomaterials **28**, 869–876 (2007)
11. Dixit, V., Van den Bossche, J., Sherman, D.M., Thompson, D.H., Andres, R.P.: Synthesis and grafting of thioctic acid-PEG-folate conjugates onto Au nanoparticles for selective targeting of folate receptor-positive tumor cells. Bioconjugate Chem. **17**, 603–609 (2006)
12. Elsabahy, M., Wooley, K.L.: Design of polymeric nanoparticles for biomedical delivery applications. Chem. Soc. Rev. **41**, 2545–2561 (2012)
13. Farokhzad, O.C., Cheng, J., Teply, B.A., Sherifi, I., Jon, S., Kantoff, P.W., Richie, J.P., Langer, R.: Targeted nanoparticle-aptamer bioconjugates for cancer chemotherapy in vivo. Proc. Natl. Acad. Sci. U.S.A. **103**, 6315–6320 (2006)
14. Fortina, P., Kricka, L.J., Surrey, S., Grodzinski, P.: Nanobiotechnology: the promise and reality of new approaches to molecular recognition. Trends Biotechnol. **23**, 168–173 (2005)

15. Freitas Jr., R.A.: The future of nanofabrication and molecular scale devices in nanomedicine. Stud. Health Technol. Inf. **80**, 45–59 (2002)
16. Ghosh Chaudhuri, R., Paria, S.: Core/shell nanoparticles: classes, properties, synthesis mechanisms, characterization, and applications. Chem. Rev. **112**, 2373–2433 (2011)
17. Groneberg, D.A., Rabe, K.F., Fischer, A.: Novel concepts of neuropeptide-based therapy: Vasoactive intestinal polypeptide and its receptors. Eu. J. Pharmacology **533**, 182–194 (2006)
18. Grady, W.M.: Epigenetic events in the colorectum and in colon cancer. Biochem. Soc. Trans. **33**, 684–688 (2005)
19. Gu, F.X., Karnik, R., Wang, A.Z., Alexis, F., Levy-Nissenbaum, E., Hong, S., Langer, R.S., Farokhzad, O.C.: Targeted nanoparticles for cancer therapy. Nano Today **2**, 14–21 (2007)
20. Gupta, P.K., Hung, C.T., Lam, F.C., Perrier, D.G.: Albumin microspheres. III. Synthesis and characterization of microsphere containing adriamycin and magnetite. Int. J. Pharm. **43**, 167–177 (1988)
21. Hattori, Y., Maitani, Y.: Enchanced in vitro DNA transfection efficiency by novel folate-linked nanoparticles in human prostate cancer and oral cancer. J. Controlled Release **97**, 173–183 (2004)
22. Hattori, Y., Maitani, Y.: Folate-linked lipid-based nanoparticle for targeted gene delivery. Curr. Drug Delivery **2**, 243–252 (2005)
23. Harris, J.M., Martin, N.E., Modi, M.: Pegylation: a novel process for modifying pharmacokinetics. Clin. Pharmacokinet. **40**, 539–551 (2001)
24. Janib, S.M., Moses, A.S., MacKay, J.A.: Imaging and drug delivery using theranostic nanoparticles. Adv. Drug. Deliv. Rev. **62**, 1052–1063 (2010)
25. Jordan, A., Scholz, R., Maier-Hauff, K., Johannsen, M., Wust, P., Nadobny, J., Schirra, H., Schmidt, H., Deger, S., Loening, S.A., Lanksch, W., Felix, R.: Presentation of a new magnetic field therapy system for the treatment of human solid tumors with magnetic fluid hyperthermia. J. Magn. Magn. Mater. **225**, 118–126 (2001)
26. Kilic, A.C., Capan, Y., Vural, I., Gursoy, R.N., Dalkara, T., Cuine, A., Hincal, A.A.: Preparation and characterization of PLGA nanospheres for the targeted delivery of NR2B-specific antisense oligonucleotides to the NMDA receptors in the brain. J. Microencapsulation **22**, 633–641 (2005)
27. Kim, T., Hyeon, T.: Applications of inorganic nanoparticles as therapeutic agents. Nanotechnology **25**, 012001 (2014)
28. Kohler, N., Sun, C., Wang, J., Zhang, M.: Methotrexate-modified superparamagnetic nanoparticles and their intracellular uptake into human cancer cells. Langmuir **21**, 8858–8864 (2005)
29. Labhasetwar, V.: Nanotechnology for drug and gene therapy: the importance of understanding molecular mechanisms of delivery. Curr. Opin. Biotechnol. **16**, 674–690 (2005)
30. Lin, S.Y., Chen, K.S., Teng, H.H., Li, M.J.: In vitro degradation and dissolution behaviours of microspheres prepared by three low molecular weight polyesters. J. Microencapsulation. **17**, 577–586 (2000)
31. Manchester, M., Singh, P.: Virus-based nanoparticles (Vnanoparticles): platform technologies for diagnostic imaging. Adv. Drug Deliv. Rev. **58**, 1505–1522 (2006)
32. Mansouri, S.P.L., Corsi, K., Benderdour, M., Beaumont, E., Fernandes, J.C.: Chitosan-DNA nanoparticles as non-viral vectors in gene therapy: strategies to improve transfection efficacy. Eur. J. Pharm. Biopharm. **57**, 1–8 (2004)
33. Moghimi, S.M., Szebeni, J.: Stealth liposomes and long circulating nanoparticles: critical issues in pharmacokinetics, opsonization and protein-binding properties. Prog. Lipid Res. **42**, 463–478 (2003)
34. Moghimi, S.M., Hunter, A.C., Murray, J.C.: Long-circulating and target-specific nanoparticles: theory to practice. Pharmacol. Rev. **53**, 283–318 (2001)
35. Nobs, L., Buchegger, F., Gurny, R.: Allémann E. Poly (lactic acid) nanoparticles labeled with biologically active Neutravidin™ for active targeting. Eur. J. Pharm. Biopharm. **58**, 483–490 (2004)

36. Nobs, L., Buchegger, F., Gurny, R., Allémann, E.: Biodegradable nanoparticles for direct or two-step tumor immunotargeting. Bioconjug Chem. **17**, 139–145 (2006)
37. Park, W., B-c, Bae, Kim, Y.-H., Na, K.: Cancer cell specific targeting of nanogels from acetylated hyaluronic acid with low molecular weight. Eur. J. Pharm. Sci. **40**, 367–375 (2010)
38. Pastorin, G., Wu, W., Wieckowski, S., Briand, J.P., Kostarelos, K., Prato, M., Bianco, A.: Double functionalisation of carbon nanotubes for multimodel drug delivery. Chem. Commun. **11**, 1182–1184 (2006)
39. Pison, U., Welte, T., Giersing, M., Groneberg, D.A.: Nanomedicine for respiratory diseases. Eur. J. Pharmacol. **533**, 341–350 (2006)
40. Panyam, J., Labhasetwar, V.: Biodegradable nanoparticles for drug and gene delivery to cells and tissue. Adv. Drug Deliv. Rev. **55**, 329–347 (2003)
41. Park, J.W.: Liposome-based drug delivery in breast cancer treatment. Breast Cancer Res. **4**, 95–99 (2002)
42. Padilla De Jesus, O.L., Ihre, H.R., Gagne, L., Frechet, J.M., Szoka Jr., F.C.: Polyester dendritic systems for drug delivery applications: in vitro and in vivo evaluation. Bioconjugate Chem. **13**, 453–461 (2002)
43. Quintana, A., Raczka, E., Piehler, L., Lee, I., Myc, A., Majoros, I., Patri, A.K., Thomas, T., Mule, J., Baker Jr., J.R.: Design and function of a dendrimer-based therapeutic nanodevice targeted to tumor cells through the folate receptor. Pharm. Res. **19**, 1310–1316 (2002)
44. Rawat, M., Singh, D., Saraf, S., Saraf, S.: Nanocarriers: promising vehicle for bioactive drugs. Biol. Pharm. Bull. **29**, 1790–1798 (2006)
45. Singh, P., Destito, G., Schneemann, A., Manchester, M.: Canine parvovirus-like particles, a novel nanomaterial for tumor targeting. J. Nanobiotechnol. **4**, 2 (2006)
46. Saul, J.M., Annapragada, A.V., Bellamkonda, R.V.: A dual ligand approach for enhancing targeting selectivity of therapeutic nanocarriers. J. Controlled Release **114**, 277–287 (2006)
47. Samlowski, W.E., McGregor, J.R., Jurek, M., Baudys, M., Zentner, G.M., Fowers, K.D.: ReGel polymer-based delivery of interleukin-2 as a cancer treatment. J. Immunother. **29**, 524–535 (2006)
48. Schatzlein, A.G.: Delivering cancer stem cell therapies-a role for nanomedicines? Eur. J. Cancer **42**, 1309–1315 (2006)
49. Sengupta, S., Eavarone, D., Capila, I., Zhao, G., Watson, N., Kiziltepe, T., Sasisekharan, R.: Temporal targeting of tumor cells and neovasculature with a nanoscale delivery system. Nature **436**, 568–572 (2005)
50. Stylios, G.K., Giannoudis, P.V., Wan, T.: Applications of nanotechnologies in medical practice. Injury **36**, 6–13 (2005)
51. Sahoo, S.K., Ma, W., Labhasetwar, V.: Efficacy of transferrin conjugated paclitaxel-loaded nanoparticles in a murine model of prostate cancer. Int. J. Cancer. **112**, 335–340 (2004)
52. Shiku, H., Wang, L., Ikuta, Y., Okugawa, T., Schmitt, M., Gu, X., Akiyoshi, K., Sunamoto, J., Nakamura, H.: Development of a cancer vaccine: peptides, proteins, and DNA. Cancer Chemother. Pharmacol. **46**, 77–82 (2000)
53. Sudimack, J., Lee, R.J.: Targeted drug delivery via the folate receptor. Adv. Drug Deliv. Rev. **41**, 147–162 (2000)
54. Shiah, J.J., Sun, Y., Peterson, C.M., Kopecek, J.: Biodistribution of free and N-(2-hydroxypropyl) methacrylamide copolymer-bound mesochlorin e(6) and adriamycin in nude mice bearing human ovarian carcinoma OVCAR-3 xenografts. J. Controlled Release **61**, 145–157 (1999)
55. Schroeder, U., Sommerfeld, P., Ulrich, S., Sabel, B.A.: Nanoparticle technology for delivery of drugs across the blood-brain barrier. J. Pharm. Sci. **87**, 1305–1307 (1998)
56. Torchilin, V.P.: Lipid-core micelles for targeted drug delivery. Curr. Drug Delivery **2**, 319–327 (2005)
57. Visaria, R.K., Griffin, R.J., Williams, B.W., Ebbini, E.S., Paciott, G.F., Song, C.W., Bischof, J.C.: Enhancement of tumor thermal therapy using gold nanoparticle-assisted tumor necrosis factor-alpha delivery. Mol. Cancer Ther. **5**, 1014–1020 (2006)

58. Vicent, M.J., Duncan, R.: Polymer conjugates: nanosized medicines for treating cancer. Trends Biotechnol. **24**, 39–47 (2006)
59. Wang, A.Z., Langer, R., Farokhzad, O.C.: Nanoparticle delivery of cancer drugs. Annu. Rev. Med. **63**, 185–198 (2012)
60. Wisse, E., Braet, F., Luo, D., Zanger, R.D., Jans, D., Crabbe, E., Vermoesen, A.N.: Structure and function of sinusoidal lining cells in the liver. Toxicol. Pathol. **24**, 100–111 (1996)
61. Yu, S., Chow, G.M.: Carboxyl group (-CO_2H) functionalized ferromagnetic iron oxide nanoparticles for potential bio-applications. J. Mat. Chem. **14**, 2781–2786 (2004)
62. Yuan, F., Dellian, M., Fukumura, D., Leunig, M., Berk, D.A., Torchilin, P., Jain, R.K.: Vascular permeability in a human tumor xenograft: molecular size dependence and cutoff size. Cancer Res. **55**, 3752–3756 (1995)

Part II
EMR Characterization of Nanoparticles

EMR of Metallic Nanoparticles

Siavash Iravani

Abstract In this chapter, current trends and future prospects about the important applications of electron magnetic resonance (EMR) spectroscopy for characterization of metallic nanoparticles are discussed.

Keywords Electron magnetic resonance · EMR · ESR · EPR · Metallic nanoparticles

1 Introduction

Nanoscience and nanotechnology have a particular field for research and development. Metal and metal oxide nanoparticles (NPs) show unique size and shape-dependent properties which are of interest for electronic, optoelectronic, biological, medical, biomedical, pharmaceutical, cosmetic, wireless electronic logic and catalytic structural applications [8, 16, 76]. Furthermore, metallic NPs have many important applications in different scientific fields, such as medical imaging, photocatalysts, nanocomposites, magnetic resonance imaging (MRI) contrast agents, filters, drug delivery systems, biosensor materials, composite fibers, electroconductive coatings, cryogenic superconducting materials, cosmetic products, electronic components, thermal spray coatings, quantum optical devices, hyperthermia of tumors, cancer therapy and diagnosis, and solar cells [65, 40, 53, 50].

Generally, electron magnetic resonance (EMR), also known as electron spin resonance (ESR), is a suitable analysis method to study magnetic properties and magnetization dynamics. EMR spectroscopy deals with microwave region of electromagnetic radiation, and it is based on the interaction between electronic magnetic moments and magnetic field. Electron Paramagnetic/Spin Resonance (EPR/ESR) is one of the most important magnetic resonance techniques, and it can

S. Iravani (✉)
Faculty of Pharmacy and Pharmaceutical Sciences,
Isfahan University of Medical Sciences, Isfahan, Iran
e-mail: siavashira@gmail.com

A.K. Shukla (ed.), *EMR/ESR/EPR Spectroscopy for Characterization of Nanomaterials*, Advanced Structured Materials 62,
DOI 10.1007/978-81-322-3655-9_4

be performed on any sample which has unpaired electron spins. Actually, EPR is the resonance absorption of electromagnetic radiation by paramagnetic substances in constant and variable magnetic fields. It measures the energy difference between energy levels for the unpaired electron in the magnetic field and gives a single line spectrum at the energy level ΔE. EPR is an analysis technique which allows researchers to study species possessing electrons with unpaired spins. Actually, EPR/ESR has proven an essential tool for study of impurity states, molecular clusters, antiferromagnetic, ferromagnetic and thin film compounds, natural or induced radicals, electron spin-based quantum information devices, optically excited paramagnetic states, and transition-metal based catalysts. Moreover, it can be applied in structural and dynamical studies of metallo-proteins, spin-labeled proteins and other complex biomolecules and their synthetic models. In summary, EMR spectra are usually characterized by the spin Hamiltonian parameters, and therefore rigorous EMR analysis requires quantum mechanical concepts. In addition, EMR can be used in order to analysis of the samples with paramagnetic centers which are the sites with unpaired electrons. Important parameters of EMR spectrum including intensity, line width, *g* value, and hyperfine splitting constant provide researchers different information about the paramagnetic specimen [59]. EMR can also provide spectroscopic information on unpaired electronic spin states at both the atomic (conventional EMR) and the nanoscale (ferromagnetic resonance), and data obtained from EMR spectrum depends upon the nature and form of specimen [61]. Among many researches on magnetic NPs, there are a considerable number of studies performed by means of EMR [47, 56, 71]. In this chapter, current trends and future prospects about the important applications of EMR spectroscopy for characterization of metallic NPs are briefly discussed.

2 Nanomaterial Properties and EMR

Nanostructured materials are defined as materials with primary particles less than 100 nm in length in at least one dimension. Nanomaterials can have size-dependent properties which result in unique behavior relative to materials composed of larger particles. Actually, on the nanoscale, electronic, optical and magnetic properties can be size dependent. Moreover, as particle size decreases, the surface/volume ratio of a material increases. Nanostructures can have high surface/volume ratios, with a large percent of total atoms present as surface atoms. Therefore, thermodynamic stability is size dependent, with surface free energy becoming an increasingly significant contribution to the total free energy as particle size decreases. The primary particle size of a material can influence surface properties, including surface area, arrangement of surface atoms, surface electronic structure, surface composition and functionality, as well as bulk properties, including shape, size, phase, electronic structure and crystallinity.

Nowadays, different properties obtained from nanomaterials lead to their technological importance, and therefore, the preparation and characterization of them are very important for scientific researches. The magnetic properties observed on a macroscopic scale are due to a very large number of atoms and, therefore, are inexistent or very different on a microscopic scale for a single atom or molecule or for a cluster of a few atoms. When fine magnetically ordered NPs are dispersed in a diamagnetic matrix, a specific type of magnetic behavior, called superparamagnetism is observed. EMR of NPs in ferrofluids, glasses, and other superparamagnetic systems has been widely studied by researchers [28, 46, 61, 27, 59]. Consequently, comprehending this phenomenon made a very important contribution to the fundamentals of magnetism and laid the foundation for the development of new materials for high-density information storage [35]. The physical properties of magnetic NPs are determined by both magnetic nature and morphologies (size and shape). A variety of experimental techniques including static magnetic, optical, magneto optical, rheological measurements, Mössbauer spectroscopy, electron microscopy, X-ray diffraction, small-angle neutron scattering have been used for determination of the magnetic characteristics of these NPs in different superparamagnetic systems and to evaluate their size distribution [35]. A tailored value of μ, even of negative μ, is obtained using magnetic resonance. Structural magnetic resonance with split-ring resonators, which are made of a nonmagnetic metal such as Cu, is a well-known technique to achieve negative μ in microwave regions [52, 62]. Moreover, an alternative route using intrinsic magnetic resonance of ferromagnetic metals, e.g., Fe, Co and Ni, has been plotted [15]. Using EMR in the metals, negative μ is obtained around an EMR frequency [13]. The resonance frequency tunability with an external magnetic field is expected to be a significant advantage of this route. It is, however, necessary to miniaturize the ferromagnetic metals, i.e., prepare ferromagnetic-metal nanoparticle systems, in order to suppress the eddy current losses. Magnetic interactions between NPs strongly affect the position and linewidth of the EMR spectra [9, 20, 32, 68]. Actually, in nanoparticle systems, two kinds of interactions have been considered: long range magnetic dipole interactions and short range direct/indirect exchange interactions [5, 4, 39]. The relative importance of each to EMR depends on the particle diameter (d) and center-to-center interparticle separation (r). Therefore, from an experimental point of view, NPs with independently controlled d and r are necessary to study EMR in an interacting nanoparticle system [30, 54].

3 EMR of Metallic NPs

3.1 Gold NPs

Gold NPs have found potential applications in many fields such as drug and gene delivery, biological detection of pathogens, catalysis of oxidation reactions, low temperature oxidation of carbon monoxide (CO), gene delivery and drug delivery systems, fluorescent biological label, animal viruses transport with gold templates, sensors (saccharides sensor and glucose oxidation), cancer therapy and diagnosis, and probing of DNA structure [2, 3, 6, 7]. Oligonucleotide-capped gold NPs have been applied for polynucleotide or protein detection using various detection and characterization techniques, such as atomic force microscopy (AFM), gel electrophoresis, scanometric assay, surface plasmon resonance imaging, amplified voltammetric detection, chronocoulometry, and Raman spectroscopy [10, 63]. Furthermore, gold NPs have been employed in immunoassay [43], protein assay [66], cancer nanotechnology [45], and capillary electrophoresis [70]. After cellular uptake, gold NPs can act as precise and powerful heaters (thermal scalpels) to kill cancer, and also they can be used as useful markers for biological screening tests [21, 57]. These NPs are capable of inducing apoptosis in β cell-chronic lymphocytic leukemia [48].

Many scientific studies have been done to monitor nanostructures and NPs for possible toxic and/or carcinogenic effects, as well as to ascertain the mechanisms of their biotoxicity. Even materials previously considered to be biologically inert should be retested when their particulate size approaches the nanoscale. In one study, the effects of particle size distribution, concentration, and agglomeration have been evaluated. Gold, which is regarded as one of the most chemically inert elements, can become catalytically active when employed in a nanoparticulate form. In one study, Ionita et al. [27] reported the air-oxidation of organic substrates containing active hydrogen atoms (e.g., amines and phosphine oxides) by phosphine and amine protected gold NPs, whereas NPs protected by more strongly bound ligands (for example, thiols) were inactive in these reactions. Overall, it was concluded that the mechanisms of such reactions could be best studied with EPR spectroscopy and available spin trapping methods [27]. Moreover, in one study, Ionita et al. [28] reported characterization of ligand dynamics attached to gold NPs. As a result, EMR spectra of a series of spin-labeled NPs at variable temperature and at different frequencies were recorded. Series of gold NPs protected by a monolayer of organic ligands and labeled them with disulfide functionalized nitroxides. The chain length of the spin labels and the surrounding ligands was systematically varied. Authors of this study found that the rotational diffusion rate of the spin labels increased with the length of the linker connecting the nitroxide unit to the gold surface.

3.2 Silver NPs

Silver NPs are important because of their extensive applications in integrated circuits [37], sensors [12], biolabeling [12], filters [12], antimicrobial deodorant fibers [75], cell electrodes [34], low-cost paper batteries (silver nanowires) (Hu et al. 2009) and antimicrobials [14]. These NPs can be used in dental materials, burn treatments, coating stainless steel materials, textile fabrics, cosmetics and sunscreen lotions [19]. Furthermore, silver NPs have been used in different fields of medicine, various industries, animal husbandry, packaging, accessories, cosmetics, health and military because of their unique antimicrobial properties [58, 19, 55, 64]. Silver NPs provide potential antimicrobial effects against infectious organisms, including *Escherichia coli*, *Bacillus subtilis*, *Vibria cholera*, *Pseudomonas aeruginosa*, *Syphilis typhus*, and *S. aureus* [14], and they were applied in disinfecting medical devices and home appliances to water treatment [29, 42, 58, 19]. For instance, plastic catheters which were coated with silver NPs had significant antimicrobial effects against *E. coli*, Enterococcus, *S. aureus*, *C. albicans*, Staphylococci, and *P. aeruginosa* [55]. These can be very useful for reducing the risk of infectious complications in patients with indwelling catheters. Furthermore, silver NPs/clay was reported to have significant antimicrobial effects against dermal pathogens, including *S. aureus, P. aeruginosa*, and *Streptococcus pyrogens,* as well as the methicillin- and oxacillin-resistant *S. aureus* [64].

In one EPR study, silver NPs were prepared by the sol-gel technique in amorphous SiO_2 and crystalline TiO_2 matrices [46]. The spin-lattice relaxation (SLR) times were investigated by pulsed electron paramagnetic resonance (EPR) spectroscopy in a temperature range between 4 and 300 K. As a result, the spin echo recoveries were slow enough, and demonstrated a biexponential character. This investigation demonstrated that for metallic silver NPs in the SiO_2 matrix, the temperature dependence of T1 can be described by the relation $(1/T1) \propto Tn$, where $0.4 < n < 1$. Contrary to this behavior, the temperature dependence of T1 has a Raman-type character ($1/T1 \propto T2$) for silver NPs in the TiO_2 matrix. Because of the amorphous phase of the SiO_2 matrix, the SiO_2 samples showed unusual behaviors.

3.3 *Nickel NPs*

Nickel NPs show unique properties (e.g., temperature and size dependence properties), and have potential applications as anode of solid oxide fuel cells or conductive electrolytic layer of proton exchange membrane fuel cells, magnetic fluid and catalyst, propellant and sintering additive, automotive catalytic converters, and supercapacitor electrode material. Moreover, these NPs are applied in coatings, plastics, nanowires, nanofibers and textiles materials [69, 26, 23, 33, 65, 44, 72, 73, 31].

EPR investigations on nickel NPs have been mostly conducted for nickel oxide and supported nickel, but its application to unsupported nickel NPs is not well-documented [1]. In one study, EMR in ferromagnetic-metal nanoparticle systems has been studied. Nickel NPs (approximately 8 nm) were inserted in polymer films. Interestingly, the EMR signal shifted and broadened, when the average distance between the NPs was decreased. Theoretical analyses based on micromagnetics simulation demonstrated that the shift of the signal was traced back to an increase in the magnetic dipole field in the nanoparticle systems due to the decrease in interparticle distance. Furthermore, the simulation showed that the perpendicular component of the dipolar field caused the broadening of the signal.

As a result, this investigation demonstrated that a dynamic analysis of the magnetization, with an explicit treatment of the magnetic dipole interactions, is essential for complete understanding of the EMR and magnetic permeability of interacting nanoparticle systems [47]. Furthermore, Alonso et al. [1] reported the characterization of the nickel NPs, utilized in transfer hydrogenation reactions with isopropanol as the hydrogen donor, as well as a series of catalytic and kinetic experiments. Presence of nickel NPs in the zero-valence state had been confirmed by EMR experiments. In another study, researches on spin dynamics in nickel NPs (3.8, 11.7, 15 and 21 nm) embedded in an amorphous SiO_2 matrix of composition 15/85 (Ni/SiO_2) were described using EMR spectroscopy at 9.28 GHz. As a result, three resonance lines were observed that EMR parameters (linewidth (ΔH), resonance field (H_r) and intensity (I_o)) were measured carefully from 5 to 300 K. Line 1 with temperature-independent $\Delta H = 50$ Oe and $g \approx 2$, and intensity varying as $1/T$, was shown to result from paramagnetic defects in the SiO_2 matrix. Lines 2 and 3, with $g \approx 2.2$ and 8, respectively, and temperature-dependent EMR parameters were assigned to nickel NPs. While line 2 with $g \approx 2.2$ was due to the majority of nickel NPs, the source of line 3 was discussed in terms of two possibilities including large clusters of blocked nickel NPs, and the inherent part of the composite asymmetric line made up of lines 2 and 3 predicted by the Raikher–Stepanov (RS) model for dispersed ferromagnets. The temperature dependence of ΔH (full width at half-maximum) of the composite line obtained by integration of the EMR spectra decreased with the increase in temperature, reaching a minimum near 300 K in agreement with the RS model. The observed decreasing asymmetry of the composite absorption spectra with increasing temperature was also in agreement with the predictions of the RS model, therefore providing a satisfactory explanation for the observed temperature dependence of the EMR spectra of nickel NPs [60].

3.4 Palladium NPs

Because of high surface-to-volume ratio and high surface energy of palladium NPs, they have applications both in heterogeneous and homogeneous catalysis. In general, most important applications of palladium NPs are: (1) Catalysts or electrocatalysts, (2) Polymer membranes, (3) Sensor design application, (4) Coatings, plastics, nanofibers and textiles, and (5) Fuel cells [38, 22, 24]. In one ESR analysis, electronic structures of monodispersed palladium (Pd) NPs with the mean diameters of 25 Å corresponding to the "magic atom number" of 561 atoms (five-shell nanoparticle) and smaller than 25 Å (22 and 23 Å) was investigated. Authors of this study reported that two broad absorption peaks corresponding to $\Delta S = 1$ and 2 and a sharp one with $\Delta S = 1$ were observed in their ESR spectra. The broad and strongest spectrum with $\Delta S = 1$ might obtained from the spin on the large orbital moment state of the NPs. The relatively strong second harmonic spectrum with $\Delta S = 2$ was characteristic of the NPs, and its low-field shift compared with bulk palladium implied the existence of the effective field from the orbital moment of the electron. Moreover, the sharp ESR signal with $\Delta S = 1$ might be derived from the NPs which have the electronic structure with one s-state electron outside the closed electron shell core with the "magic atom number" [67].

3.5 *Cerium and Gadolinium Doped Nickel Ferrite NPs*

Gadolinium (Gd) NPs can be used in MRI for detection of early stage cancer (especially breast cancer order) and in order to increase the speed and capacity of computer memory. These NPs acts as hosts for X-ray cassettes and in scintillator materials for computer tomography. Furthermore, Gd NPs applied in neutron capture therapy and for MRI contrast enhancement. NPs of gadolinium oxide can be used in different areas including fluorescent materials, special optical glass, electric industry, as additives, catalysts, and dopants, neutron converters, cathode-ray tubes and UV detectors, high resolution X-ray medical imaging, ferroelectric memory, electroluminescent devices, field emission displays and plasma display panels, luminescence, scintillators and sintering aids, and solid oxide fuel cells [11, 18, 36, 51]. Cerium (Ce) oxide and free ceria salts have a number of different industrial roles. Free cerium is used in glass polishing, cracking in petroleum, fertilizers and oxygen sensors. Cerium oxide has found a number of industrial uses, especially in automotive exhausts where it has been shown to effectively reduce the amount of pollutants released from exhaust fumes [74, 49].

In one study, the EPR spectra of cerium and gadolinium doped nickel ferrite NPs were recorded from 120 to 300 K. As a result, doping with cerium and gadolinium reduced the line width and *g* value in comparison to that of pure nickel ferrite. Cerium doped samples have the lowest values of both these parameters at room temperature. This indicates that cerium doped samples show lowest loss and is suitable for high frequency devices. EPR spin numbers are reduced while the spin relaxation time is increased after doping with rare earth ions. Gadolinium doped samples have higher values of relaxation time and lower spin numbers in comparison to that of cerium doped samples [17].

4 Conclusion

In conclusion, characterization of nanostructures and analysis of their properties using appropriate techniques are important issues in nanotechnology, and therefore in order to have precise knowledge on the characteristics and properties of nanostructured materials, it is critical to fabricate monodispersed ones with a perfectly controlled size (sample uniformity in terms of size, morphology, internal structure, and surface chemistry).

Magnetic effects as one of the size-dependent physical properties show great attractive for a broad range of biomedical and pharmaceutical applications (for example, cancer diagnosis and therapy). Actually, the EMR in combination with other experimental techniques provides a powerful tool of studying the physical properties of magnetic NPs owing to its sensitivity to both the magnetic state and the morphological characteristics. The EMR in ferromagnetic-metal nanoparticle systems is not well-understood. Notably, the magnetic behavior and dynamics of magnetization in nanoparticle systems are dependent both on single-particle properties (such as magneto-crystalline or shape anisotropies of a particle) and on interparticle interactions. For magnetically isolated particles with negligible interaction, only single-particle properties need to be considered. In contrast, in the case of strongly interacting particle systems, an explicit treatment of the interactions is essential.

References

1. Alonso, F., Riente, P., Sirvent, J.A., Yus, M.: Nickel nanoparticles in hydrogen-transfer reductions: characterisation and nature of the catalyst. Appl. Catal. A: Gen. **378**, 42 (2010)
2. Ankamwar, B., Chaudhary, M., Mural, S.: Gold nanotriangles biologically synthesized using tamarind leaf extract and potential application in vapor sensing. Synth. React. Inorg. Metal-Org. Nanometal. Chem. **35**, 19–26 (2005)
3. Ankamwar, B., Damle, C., Ahmad, A., Sastry, M.: Biosynthesis of gold and silver nanoparticles using *Emblica officinalis* fruit extract, their phase transfer and transmetallation in an organic solution. J. Nanosci. Nanotechnol. **5**(10), 1665–1671 (2005)
4. Arias, R., Chu, P., Mills, D.: Dipole exchange spin waves and microwave response of ferromagnetic spheres. Phys. Rev. B **71**, 224410 (2005)
5. Arias, R., Mills, D.: Theory of collective spin-wave modes of interacting ferromagnetic spheres. Phys. Rev. B **70**, 104425 (2004)
6. Bhattacharya, R., Murkherjee, P.: Biological properties of "naked" metal nanoparticles. Adv. Drug Deliv. Rev. **60**, 128–1306 (2008)
7. Bhumkar, D.R., Joshi, H.M., Sastry, M., Pokharkar, V.B.: Chitosan reduced gold nanoparticles as novel carriers for transmucosal delivery of insulin. Pharm. Res. **24**, 1415–1426 (2007)
8. Bogunia-Kubik, K., Sugisaka, M.: From molecular biology to nanotechnology and nanomedicine. Biosystems **65**, 123–138 (2002)
9. Butera, A., Zhou, J.N., Barnard, J.: Ferromagnetic resonance in as-deposited and annealed $Fe - SiO_2$ heterogeneous thin films. Phys. Rev. B **60**, 12270 (1999)
10. Cai, W., Gao, T., Hong, H., Sun, J.: Applications of gold nanoparticles in cancer nanotechnology. Nanotechnol. Sci. Appl. **1**, 17–32 (2008)
11. Cao, C.-Y., Shen, Y.-Y., Wang, J.-D., Li, L., Liang, G.L.: Controlled intracellular self-assembly of gadolinium nanoparticles as smart molecular MR contrast agents. Sci. Rep. **3**, 1024 (2013)
12. Cao, G.: Nanostructures and Nanomaterials: Synthesis, Properties and Applications. Imperial College Press, london (2004)
13. Chikazumi, S.: Physics of Ferromagnetism. Oxford University Press, Oxford (1997)
14. Cho, K.-H., Park, J.-E., Osaka, T., Park, S.G.: The study of antimicrobial activity and preservative effects of nanosilver ingredient. Electrochim. Acta **51**, 956–960 (2005)

15. Chui, S.T., Hu, L.: Theoretical investigation on the possibility of preparing left-handed materials in metallic magnetic granular composites. Phys. Rev. B **65**, 144407 (2002)
16. Daniel, M.C., Astruc, D.: Gold nanoparticles: assembly, supramolecular chemistry, quantum-size-related properties, and applications toward biology, catalysis, and nanotechnology. Chem. Rev. **104**, 293–346 (2004)
17. Dixit, G., Singh, J.P., Srivastava, R.C., Agrawal, H.-M.: Magnetic resonance study of Ce and Gd doped $NiFe_2O_4$ nanoparticles. J. Magn. Magn. Mater. **324**, 479–483 (2012)
18. Duc, G.L., Roux, S., Paruta-Tuarez, A., Dufort, S., Brauer, E., Marais, A., Truillet, C., Sancey, L., Perriat, P., Lux, F., Tillement, O.: Advantages of gadolinium based ultrasmall nanoparticles vs molecular gadolinium chelates for radiotherapy guided by MRI for glioma treatment. Cancer Nanotechnol. **5**, 4 (2014)
19. Duran, N., Marcato, D.P., De Souza, H.I., Alves, L.O., Espsito, E.: Antibacterial effect of silver nanoparticles produced by fungal process on textile fabrics and their effluent treatment. J. Biomed. Nanotechnol. **3**, 203–208 (2007)
20. Ebels, U., Duvail, J.-L., Wigen, P.E., Piraux, L., Buda, L.D., Ounadjela, K.: Ferromagnetic resonance studies of Ni nanowire arrays. Phys. Rev. B **64**, 144421 (2001)
21. El-Sayed, I.H., Huang, X., El-Sayed, M.A.: Selective laser photo-thermal therapy of epithelial carcinoma using anti-EGFR antibody conjugated gold nanoparticles. Cancer Lett. **239**, 129–135 (2006)
22. Gao, S., Cao, M., Lia, W., Cao, R.: Palladium nanoparticles in situ generated in metal-organic films for catalytic applications. J. Mater. Chem. A **2**, 12185–12193 (2014)
23. Gubin, S.P., Koksharov, Y.A., Khomutov, G.B., Yurkov, G.Y.: Magnetic nanoparticles: preparation, structure and properties. Russ. Chem. Rev. **74**, 489 (2005)
24. Hennebel, T., De Corte, S., Verstraete, W., Boon, N.: Microbial production and environmental applications of Pd nanoparticles for treatment of halogenated compounds. Curr. Opin. Biotechnol. **23**, 555–561 (2012)
25. Hu, L, Choi, J.W., Yang, Y., Jeong, S., La Mantia, F., Cui, L.-F., Cui, Y.: Highly conductive paper for energy-storage devices. Proc. Nat. Acad. Sci. USA **106**, 21490–21494, S21490/21491–S21490/21413 (2009)
26. Hyeon T (2003) Chemical synthesis of magnetic nanoparticles. Chem. Commun. 927
27. Ionita, P., Conte, M., Gilbert, B.C., Chechik, V.: Gold nanoparticle-initiated free radical oxidations and halogen abstractions. Org. Biomol. Chem. **5**, 3504–3509 (2007)
28. Ionita, P., Wolowska, J., Chechik, V., Caragheorgheopol, A.: Ligand dynamics in spinlabeled Au nanoparticles. J. Phys. Chem. C **111**, 16717 (2007)
29. Jain, P., Pradeep, T.: Potential of silver nanoparticle-coated polyurethane foam as an antibacterial water filter. Biotechnol. Bioeng. **90**, 59–63 (2005)
30. Jensen, P.J., Pastor, G.: Scaling behavior of the dipole-coupling energy in two-dimensional disordered magnetic nanostructures. Phys. Rev. B **68**, 184420 (2003)
31. Jiang, Z., Xie, J., Jiang, D., Wei, X., Chen, M.: Modifiers-assisted formation of nickel nanoparticles and their catalytic application to p-nitrophenol reduction. CrystEngComm **15**, 560–569 (2013)
32. Jung, S., Ketterson, J.B., Chandrasekhar, V.: Micromagnetic calculations of ferromagnetic resonance in submicron ferromagnetic particles. Phys. Rev. B **66**, 132405 (2002)
33. Karmhag, R., Tesfamichael, T., Wackelgard, E., Nikalsson, G., Nygren, M.: Oxidation kinetics of nickel particles: comparison between free particles and particles in an oxide matrix. Sol. Energy **68**, 329 (2000)
34. Klaus-Joerger, T., Joerger, R., Olsson, E., Granqvist, C.G.: Bacteria as workers in the living factory: metal-accumulating bacteria and their potential for materials science. Trends Biotechnol. **19**(1), 15–20 (2001)
35. Kliava, J.: Electron magnetic resonance of nanoparticles: superparamagnetic resonance. In: Gubin, S.P. (ed.) Magnetic Nanoparticles. WILEY-VCH Verlag GmbH & Co. KGaA, Weinheim (2009)
36. Korkusuz, H., Ulbrich, K., Welzel, K., Koeberle, V., Watcharin, W., Bahr, U., Chernikov, V., Knobloch, T., Petersen, S., Huebner, F., Ackermann, H., Gelperina, S., Kromen, W.,

Hammerstingl, R., Haupenthal, J., Gruenwald, F., Fiehler, J., Zeuzem, S., Kreuter, J., Vogl, T. J., Piiper, A.: Transferrin-coated gadolinium nanoparticles as MRI contrast agent. Mol. Imaging Biol. **15**, 148–154 (2013)

37. Kotthaus, S., Gunther, B.H., Hang, R., Schafer, H.: Study of isotropically conductive bondings filled with aggregates of nano-sited Ag-particles. IEEE Trans. Compon. Packag. Technol. **20**(1), 15–20 (1997)
38. Kumar, V.K.R., Krishnakumar, S., Gopidas, K.R.: Synthesis, characterization and catalytic applications of palladium nanoparticle-cored dendrimers stabilized by metal-carbon bonds. Eur. J. Org. Chem. **2012**, 3447–3458 (2012)
39. Laroze, D., Vargas, P.: Dynamical behavior of two interacting magnetic nanoparticles. Physica B: Condens. Matter **372**, 332–336 (2006)
40. Lee, H.Y., Li, Z., Chen, K., Hsu, A.R., Xu, C., Xie, J., Sun, S., Chen, X.: PET/MRI dual-modality tumor imaging using arginine-glycine-aspartic (RGD)-conjugated radiolabeled iron oxide nanoparticles. J. Nucl. Med. **49**, 1371–1379 (2008)
41. Lee, S.I., Lee, N., Park, J., Kim, B.H., Yi, Y., Kim, T., Kim, T.K., Lee, I.H., Paik, S.R., Hyeon, T.: Ni/NiO core/shell nanoparticles for selective binding and magnetic separation of histidine tagged proteins. J. Am. Chem. Soc. **128**, 10658 (2006)
42. Li, Q., Mahendra, S., Lyon, D., Brunet, L., Liga, M., Li, D., Alvarez, P.: Antimicrobial nanomaterials for water disinfection and microbial control: potential applications and implications. Water Res. **42**, 4591–4602 (2008)
43. Liu, X., Dai, Q., Austin, L., Coutts, J., Knowles, G., Zou, J., Chen, H., Huo, Q.: A one-step homogeneous immunoassay for cancer biomarker detection using gold nanoparticle probes coupled with dynamic light scattering. J. Am. Chem. Soc. **130**, 2780–2782 (2008)
44. Llamazares, S.R., Merchan, J., Olmedo, I., Marambio, H.P., Munoz, J.P., Jara, P., Sturm, J.C., Chornik, B., Pena, O., Yutronic, N., Kogan, M.J.: Ni/Ni oxides nanoparticles with potential biomedical applications obtained by displacement of a nickel-organometallic complex. J. Nanosci. Nanotechnol. **8**, 3820 (2008)
45. Medley, C.D., Smith, J.E., Tang, Z., Wu, Y., Bamrungsap, S., Tan, W.: Gold nanoparticle-based colorimetric assay for the direct detection of cancerous cells. Anal. Chem. **80**, 1067–1072 (2008)
46. Mitrikas, G., Deligiannakis, Y., Trapalis, C.C., Boukos, N., Kordas, G.: CW and pulsed EPR study of silver nanoparticles in a SiO_2 matrix. J. Sol-Gel. Sci. Technol. **13**, 503 (1998)
47. Mitsumata, C., Tomita, S., Hagiwara, M., Akamatsu, K.: Electron magnetic resonance in interacting ferromagnetic-metal nanoparticle systems: experiment and numerical simulation. J. Phys.: Condens. Matter **22**, 016005 (2010)
48. Mukherjee, P., Bhattacharya, R., Bone, N., Lee, Y.K., Patra, C.R., Wang, S., Lu, L., Secreto, C., Banerjee, P.C., Yaszemski, M.J., Kay, N.E., Mukhopadhyay, D.: Potential therapeutic application of gold nanoparticles in β-chronic lymphocytic leukemia (BCLL): enhancing apoptosis. J. Nanobiotechnol. **5**, 4 (2007)
49. Nikolaou, K.: Emissions reduction of high and low polluting new technology vehicles equipped with a CeO2 catalytic system. Sci. Total Environ. **235**, 71–76 (1999)
50. Panigrahi, S., Kundu, S., Ghosh, S., Nath, S., Pal, T.: General method of synthesis for metal nanoparticles. J. Nanopart. Res. **6**, 411–414 (2004)
51. Park, J.Y., Baek, M.J., Choi, E.S., Woo, S., Kim, J.H., Kim, T.J., Jung, J.C., Chae, K.S., Chang, Y., Lee, G.H.: Paramagnetic ultrasmall gadolinium oxide nanoparticles as advanced T1 MRI contrast agent: account for large longitudinal relaxivity, optimal particle diameter, and in vivo T1 MR images. ACS Nano **3**, 3663–3669 (2009)
52. Pendry, J.B., Holden, A.J., Robbins, D.J., Stewart, W.J.: Magnetism from conductors and enhanced nonlinear phenomena. IEEE Trans. Microw. Theory Tech. **47**, 2075–2084 (1999)
53. Pissuwan, D., Valenzuela, S.M., Cortie, M.B.: Therapeutic possibilities of plasmonically heated gold nanoparticles. Trends Biotechnol. **24**, 62–67 (2006)
54. Puntes, V.F., Gorostiza, P., Aruguete, D.M., Bastus, N.G., Alivisatos, A.P.: Collective behaviour in two-dimensional cobalt nanoparticle assemblies observed by magnetic force microscopy. Nat. Mater. **3**, 263–268 (2004)

55. Roe, D., Karandikar, B., Bonn-Savage, N., Gibbins, B., Roullet, J.B.: Antimicrobial surface functionalization of plastic catheters by silver nanoparticles. J. Antimicrob. Chemother. **61**, 869–876 (2008)
56. Salado, J., Insausti, M., Lezama, L., de Muro, I.G., Goikolea, E., Rojo, T.: Preparation and characterization of monodisperse Fe_3O_4 nanoparticles: an electron magnetic resonance study. Chem. Mater. **23**, 2879–2885 (2011)
57. Salata, O.: Applications of nanoparticles in biology and medicine. J. Nanobiotechnol. **2**, 3 (2004)
58. Shahverdi, R.A., Fakhimi, A., Shahverdi, H.R., Minaian, S.: Synthesis and effect of silver nanoparticles on the antibacterial activity of different antibiotics against Staphyloccocus aureus and Escherichia coli. Nanomed.: Nanotechnol. Biol. Med. **3**, 168–171 (2007)
59. Shukla, A.K.: Electron magnetic resonance (EMR) technique and nanoparticle characterization. In: Lungu, M. et al. (ed.) Nanoparticles' Promises and Risks. Springer International Publishing Switzerland, pp. 235–244 (2015)
60. Singh, V., Seehra, M.S.: Temperature and size dependence of electron magnetic resonance spectra of Ni nanoparticles embedded in an amorphous SiO_2 matrix. J. Phys.: Condens. Matter **21**, 456001 (2009)
61. Smirnov, A.I.: EPR studies of nanomaterials. In: Misra, S.K. (ed.) Multifrequency Electron Paramagnetic Resonance: Theory and Applications, p. 825. Wiley, New York (2011)
62. Smith, D.R., Padilla, W.J., Vier, D.C., Nemat-Nasser, S.C., Schultz, S.: Composite medium with simultaneously negative permeability and permittivity. Phys. Rev. Lett. **84**, 4184 (2000)
63. Sperling, R.A., Gil, P.R., Zhang, F., Zanella, M., Parak, W.J.: Biological applications of gold nanoparticles. Chem. Soc. Rev. **37**, 1896–1908 (2008)
64. Su, H.L., Chou, C.C., Hung, D.J., Lin, S.H., Pao, I.C., Lin, J.H., Huang, F.L., Dong, R.X., Lin, J.-J.: The disruption of bacterial membrane integrity through ROS generation induced by nanohybrids of silver and clay. Biomaterials **30**, 5979–5987 (2009)
65. Tan, M., Wang, G., Ye, Z., Yuan, J.: Synthesis and characterization of titania-based monodisperse fluorescent europium nanoparticles for biolabeling. J. Lumin. **117**, 20–28 (2006)
66. Tang, D., Yuan, R., Chai, Y.: Biochemical and immunochemical characterization of the antigen-antibody reaction on a non-toxic biomimetic interface immobilized red blood cells of crucian carp and gold nanoparticles. Biosens. Bioelectron. **22**, 1116–1120 (2007)
67. Teranishi, T., Hori, H., Miyake, M.: ESR study on palladium nanoparticles. J. Phys. Chem. B **101**, 5774–5776 (1997)
68. Tomita, S., Hagiwara, M., Kashiwagi, T., Tsuruta, C., Matsui, Y., Fujii, M., Hayashi, S.: Ferromagnetic resonance study of diluted Fe nanogranular films. J. Appl. Phys. **95**, 8194 (2004)
69. Tseng, W., Chen, C.: Dispersion and rheology of nickel nanoparticle inks. J. Mater. Sci. **41**, 1213 (2006)
70. Tseng, W.L., Huang, M.F., Huang, Y.F., Chang, H.-T.: Nanoparticle-filled capillary electrophoresis for the separation of long DNA molecules in the presence of hydrodynamic and electrokinetic forces. Electrophoresis **26**, 3069–3075 (2005)
71. Usselman, R.J., Russek, S.E., Klem, M.T., Allen, M.A., Douglas, T., Young, M., Idzerda, Y. U., Singel, D.J.: Temperature dependence of electron magnetic resonance spectra of iron oxide nanoparticles mineralized in Listeria innocua protein cages. J. Appl. Phys. **112**, 084701 (2012)
72. Veerakumar, P., Chen, S.M., Madhu, R., Veeramani, V., Hung, C.T., Liu, S.B.: Nickel nanoparticle-decorated porous carbons for highly active catalytic reduction of organic dyes and sensitive detection of Hg(II) ions. ACS Appl. Mater. Interfaces **7**, 24810–24821 (2015)
73. Wu, X., XingW, Zhang L., Zhuo, S., Zhou, J., Wang, G., Qiao, S.: Nickel nanoparticles prepared by hydrazine hydrate reduction and their application in supercapacitor. Powder Technol. **224**, 162–167 (2012)

74. Xu, X., Zhu, W., Wang, Z., Witkamp, G.J.: Distributions of rare earths and heavy metals in field-grown maize after application of rare earth-containing fertilizer. Sci. Total Environ. **293**, 97–105 (2002)
75. Zhang, W., Wang, G.: Research and development for antibacterial materials of silver nanoparticle. New Chem. Mater. **31**(2), 42–44 (2003)
76. Zharov, V.P., Kim, J.-W., Curiel, D.T., Everts, M.: Self-assembling nanoclusters in living systems: application for integrated photothermal nanodiagnostics and nanotherapy. Nanomed. Nanotechnol. Biol. Med. **1**, 326–345 (2005)

Electron Spin Resonance Applied to Nanosized-Doped Oxides

Cesare Oliva and Marco Scavini

Abstract The electron spin resonance (ESR or EPR) is a spectroscopic technique particularly suited to investigate solid paramagnetic samples like powders composed of nanosized paramagnetic particles. In this chapter we review ESR investigations on samples of this kind. The effects of doping oxidic nanoparticles by metal cations or by nonmetal species are shown in the first paragraph. ESR can provide information on the energy of the unpaired electron and on the crystal field around it. When two or more unpaired electrons are present on a paramagnetic system, their reciprocal interactions supply important information on nature and distribution of dopant magnetic ions. Moreover, compositional fluctuations at the nanometric scale can induce the formation of extended clusters of magnetic ions. Systems of this kind are discussed in the second and third paragraph of the chapter.

Keywords ESR · Nanomaterials · Smart materials · Oxides · Doping · Superparamagnetism · Ferromagnetism

1 Introduction

One of a more interesting parameter measurable by electron spin resonance (ESR or EPR) is the *spin–spin relaxation time* T_2, which is affected by many important physical phenomena and is easily evaluated being inversely proportional to the peak-to-peak width of the first-derivative EPR line. The magnetic interaction between the unpaired electron $\boldsymbol{S}$ spin and the external magnetic field $\boldsymbol{B}$ is anisotropic in a solid matrix, so that the resonance field depends on a $\boldsymbol{g}$ "tensor,"

C. Oliva (✉) · M. Scavini
Dipartimento di Chimica dell'Università degli Studi di Milano, Milan, Italy
e-mail: cesare.oliva@unimi.it

C. Oliva · M. Scavini
CNR-Istituto Scienza e Tecnologie Molecolari, Milan, Italy

C. Oliva
Accademia Nazionale di Scienze Lettere ed Arti (Modena), Modena, Italy

A.K. Shukla (ed.), *EMR/ESR/EPR Spectroscopy for Characterization of Nanomaterials*, Advanced Structured Materials 62,
DOI 10.1007/978-81-322-3655-9_5

reducing to a vector in a proper reference frame. ***g*** provides information on the unpaired electron energy of a paramagnetic dopant and on the crystal field around it. These data add to the information obtainable from the ***a*** hyperfine *electron spin–nuclear spin* coupling constant which also has a tensorial character in the solids and in general reduces to a vector in the same reference frame which diagonalizes ***g***. These will be the main topics of the first paragraph of this Chapter. In particular, the effects of doping diamagnetic nanoparticles of TiO_2 by metal cations or by nonmetal species, or by both these kinds of dopants, will be reported, together with other interesting examples.

When two or more unpaired electrons are present on the same paramagnetic system, their reciprocal interaction can be accounted for by adding suitable spin Hamiltonian contributions, containing both the exchange J (scalar) and the dipolar ***D*** (tensorial) *electron spin–electron spin* interaction terms. Sometimes, ***D*** is conveniently represented by its "diagonal" elements D and E as in the case of the Fe^{3+} ($S = 5/2$) ions reported in the second paragraph of this Chapter.

A different situation occurs when the paramagnetic particles do not contain isolated groups of reciprocally interacting unpaired electrons set at fixed reciprocal distances but, instead, a stochastic distribution of unpaired electrons spread everywhere and all interacting to each other. In this case, a unique unresolved EPR feature is detected, which can have a Gaussian or a Lorentzian shape, depending on the prevailing of the dipolar or of the exchange interactions, respectively. Systems of this kind are discussed in the paragraphs 2 and 3. In particular, the exchange phenomena observed with chains formed of paramagnetic metal ions alternated to oxygen ions will be discussed in the third paragraph. At low temperature these systems can become diamagnetic or ferromagnetic, with the consequent decrease or increase of the spectral intensity. In the latter case, a ferromagnetic resonance (FMR) spectrum is obtained, as those reported in the paragraph 2 with clusters of Sm^{2+} and of Gd^{3+} ions. Nanosized magnetic particles can contain a single FM domain. Then, the magnetic behavior is more similar to that of a paramagnetic sample, and a superparamagnetic (SPM) system is obtained, as discussed with nanosized clusters of Fe or Gd ions.

2 Isolated Paramagnetic Ions Doping Diamagnetic Nanoparticles

2.1 Doped Titanium Dioxide

TiO_2 has been the subject of extensive research due to its relevant photochemical and photo-physical properties combined with its natural abundance, low cost, and high chemical- and photo-stability. Applications of nano-crystalline TiO_2 particles are quite numerous and span from photo-catalysis, self-cleaning, and antifogging devices, to solar energy conversion and photo-electrochemical water splitting [1–3].

However, the TiO_2 photo-activity is restricted to the UV region, which represents only a few percent of the energy of solar spectrum. This limitation is due to the wide band gap (>3 eV) characterizing this oxide. Several approaches have been developed to reduce this energy gap, aiming at utilizing also the visible light ($\lambda > 400$ nm). Two main different possibilities of doping have been proposed depending on the final application: either by metal cations or by *p*-block nonmetal species (N, C, S, P, etc.).

2.1.1 Doping Titanium Dioxide by Metal Cations

Several metals of the first transition series have been proposed as dopant of TiO_2. However, the enhanced absorption in the visible region has not always been accompanied by a significant improvement of the material photocatalytic performance. Rather, even detrimental effects were observed in some cases [4]. The doping effects are somehow different in the case of TiO_2 doped with high valence Sb(V) (d^{10}) or Nb(V) (d^0) ions, the latter being able to promote both photocatalytic activity and water splitting processes [5], even combined with other lower valence transition cations [6]. Furthermore, Nb–TiO_2 shows an electronic conductivity comparable to that of the classic transparent conductive systems (ITO, FTO), while maintaining optimal visible transparency [7] and is employed for the development of highly efficient photo-anodes in dye-sensitized solar cells showing a reduced recombination loss and increased electron injection [8, 9].

A significant band gap reduction has been observed also with Cu-doped TiO_2 nanoparticles [10].

The EPR spectroscopy is an important tool to determine the valence state of a Cu-dopant ion, as only Cu^{2+} is paramagnetic. Furthermore, the EPR spectrum provides g values which are very sensitive to the coordination environment of the host framework, through the equations

$$g_{//} = g_e + 8\frac{\lambda}{\Delta_2} \tag{1}$$

$$g_{\perp} = g_e + 2\frac{\lambda}{\Delta_1} \tag{2}$$

valid for tetragonally distorted octahedral crystal field, or other analogous equations reported in literature for other kinds of crystal fields [11]. The EPR spectrum of a Cu-doped TiO_2 is like that reported in Fig. 1. This is composed of two rather well-resolved spectral regions centered at $g_{//} = 2.33$ and $g_{\perp} = 2.08$, respectively, further split into four lines, well resolved in the $g_{//}$ region only, due to the hyperfine coupling between unpaired electron and Cu nucleus (which has spin $I = 3/2$). The spectrum of Fig. 1 is perfectly compatible with Eqs. (1) and (2), i.e., with a Cu^{2+} ion in an octahedral (tetragonally distorted) field as that of Fig. 2b. Therefore, the Cu^{2+} is substituted for a Ti^{4+} on the lattice site, being placed in the octahedral

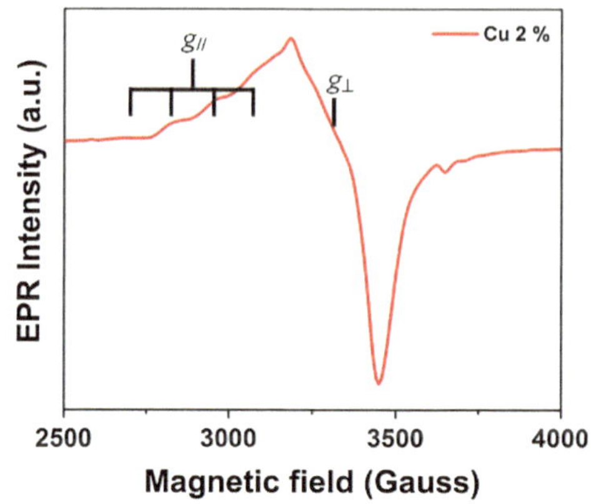

Fig. 1 Room temperature EPR spectrum of Cu-2 %-doped TiO_2 nanoparticles. Adapted from Ref. [10]

coordination of TiO_2. Each of these substitutions was generating an oxygen vacancy (Fig. 2a) and was creating a distortion of the octahedral field. Furthermore, the spectral lines are rather broad, suggesting that the Cu^{2+} ions are dipolarly interacting to each other. This could be associated with formation of copper-based nano-clusters, since bulk copper clusters were not noticed in the diffraction patterns. Therefore, charge transitions from the valence band of TiO_2 to the Cu^{2+} nano-clusters attached to TiO_2 were hypothesized. These electron transitions would add to the *d–d* ones of the Cu^{2+} ions present in the crystalline environment of TiO_2.

2.1.2 Doping Titanium Oxide by Nonmetal Species

N has been probably the more deeply investigated among the nonmetal dopants of TiO_2 and many EPR studies have been published on this subject [3, 12–30].

The chemical nature of the TiO_2-doping nitrogen species has been discussed by Livraghi et al. [16] and by Di Valentin et al. [17] and represented as in Figs. 3 and 4.

Nitrogen atoms can be substitutional of an oxygen atom (a) or belonging to an interstitial NO molecule (b) (Fig. 3). In both these cases N doping creates localized intra-gap electronic levels (Fig. 4). Therefore, the "collective" $N_b^{\bullet}$ symbol was proposed [19] for these bulk nitrogen species, further labeled by "i" or by "s" in the cases in which they are unequivocally "interstial" or "substitutional." It has been reported that different sample preparation procedures and N sources lead to samples characterized by different activities and durabilities [3, 18, 29].

This matter has been reviewed in part by Nick Serpone [20] and by Emeline et al. [24], who added some interesting considerations on possible light-absorbing mechanisms, competitive to that above mentioned for the N doping (Fig. 5d) but, instead, attributable to the formation of Ti^{3+} and to color centers F^+ (Fig. 5e), not able to excite electrons up to the conduction band of the sample. F^+ centers could be formed through the process

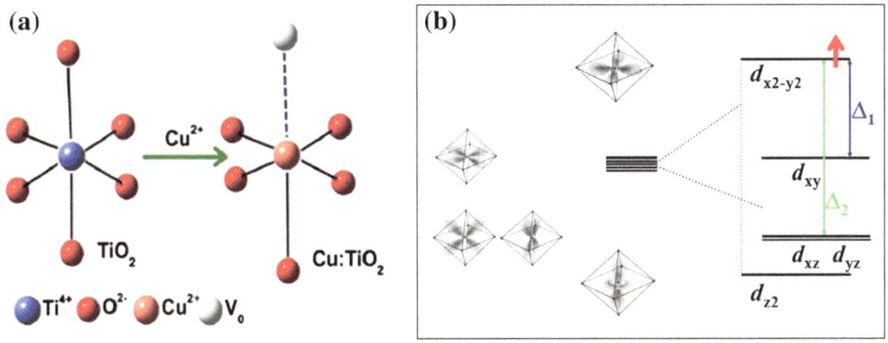

Fig. 2 **a** Distortion of the TiO_6 octahedra on doping Cu^{2+} for Ti^{4+} in a lattice site. An oxygen vacancy VO (white ball) is generated nearby Cu^{2+}. Adapted from Ref. [10]. **b** The $3d^9$ electron energy levels of Cu^{2+} in an octahedral (tetragonally distorted) field. The reported example is the extreme one of square planar fields, in which d_z2 is at a very low-energy value. All the represented levels are doubly occupied, with the exception of $d_x2 -_y2$, which is singly occupied

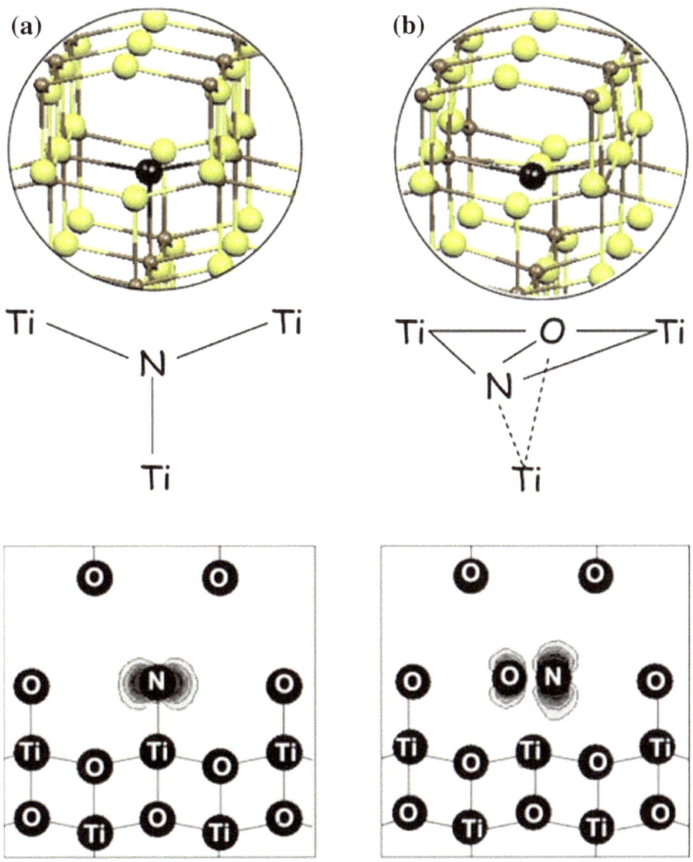

Fig. 3 N (**a**) and ON (**b**), substitutionally and interstitially doping TiO_2. Reprinted with permission from Ref. [17]. Copyright (2005) American Chemical Society

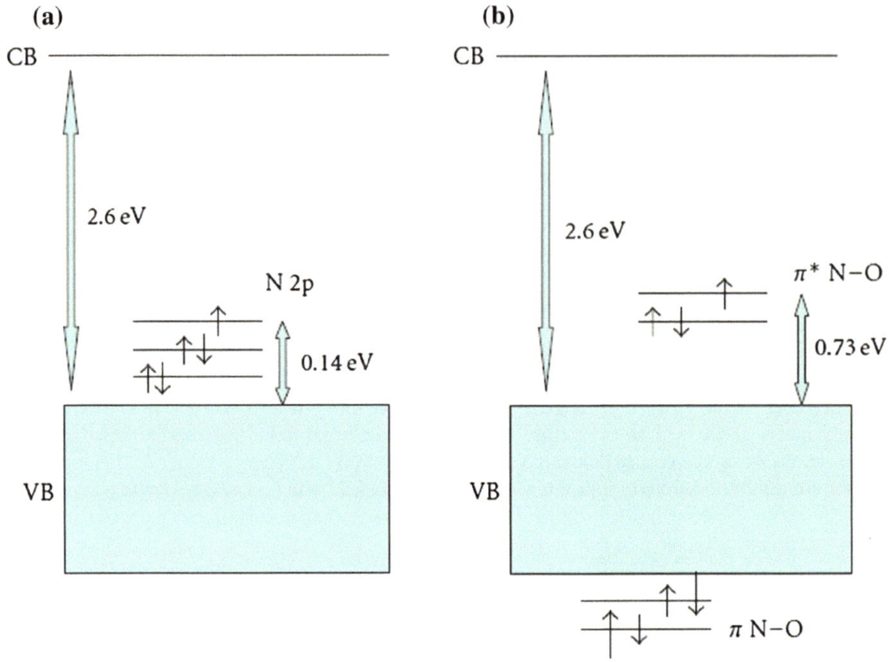

Fig. 4 Intra-gap electron levels introduced by N (**a**) and by ON (**b**), respectively, substitutionally and interstitially doping TiO_2. Reprinted with permission from Ref. [24]

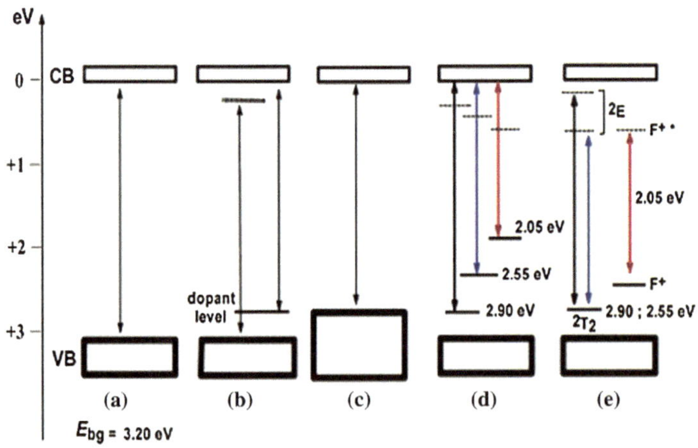

Fig. 5 Various schemes illustrating the possible changes that might occur to the electron energy band gap of anatase TiO_2 on doping with various nonmetals: (**a**) band gap of pristine TiO_2; (**b**) doped TiO_2 with localized dopant levels near the VB and the CB; (**c**) band gap narrowing resulting from broadening of the VB; (**d**) localized dopant levels and electronic transitions to the CB; and (**e**) electronic transitions from localized levels near the VB to their corresponding excited states for Ti^{3+} and F^{+} centers. Reprinted with permission from Ref. [20]. Copyright (2006) American Chemical Society

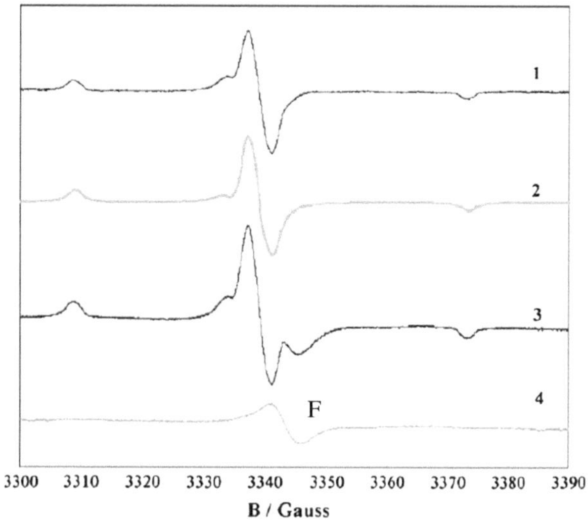

Fig. 6 EPR spectra of N–TiO_2. Initial N/Ti atomic ratio: 0.10, experimental (*1*); 0.10, simulated with $g_x = 2.0066$; $g_y = 2.0054$; $g_z = 0\ 2.0040$; $A_x \cong A_y \cong 0$; $A_z \cong 32.2$ G (*2*); ≥ 0.20 experimental (*3*); difference track 3 minus track 2 (*4*), resulting in a Lorentzian-shaped line attributed to F^+ defects. Adapted with permission from Ref. [26]

$$V_O^{\bullet\bullet} + 1e^- \rightarrow F^+ \tag{3}$$

in which $V_O^{\bullet\bullet}$ indicates an electron pair-deficient oxygen vacancy (sometimes indicated also as F^{++} center) becoming F^+ when trapping an electron.

The EPR line of the paramagnetic F^+ defects is observable even at room temperature at enough great amounts of N doping, as shown in Fig. 6 (line F).

The concentration of the F^+ defects seems connected to the loss of activity occurring during sample irradiation. The intensity of the F line increases also with the aging of the sample and decreases with its regeneration by oxidation [26].

Preparation procedures and N sources control also the material microstructure [31], which in turn determines the concentration and nature of both extrinsic and intrinsic point defects. As for the latter, it has been hypothesized that a F^+ center might provide its electron to a neighboring Ti^{4+} ion to reduce it to Ti^{3+} and determine the formation of new gap states just below the conduction band (CB) [15, 24, 32].

2.1.3 Co-doping Titanium Oxide by Nonmetal and Metal Species

Donor mid-gap states are usually quite deep within the gap, and they may be responsible for a faster electron–hole recombination time [33]. Compensating titania with nonmetals and transition metals is a highly promising way to obtain second-generation photo-catalysts, where the apparent band gap decrease occurs without increasing the electron–hole recombination rate [34–38]. Marquez studied the structural and electronic properties of W–N codoped titania observing enhanced photocatalytic properties [39]. Significant results have been reported also in the case of V–N and Fe–N titania systems [40, 41].

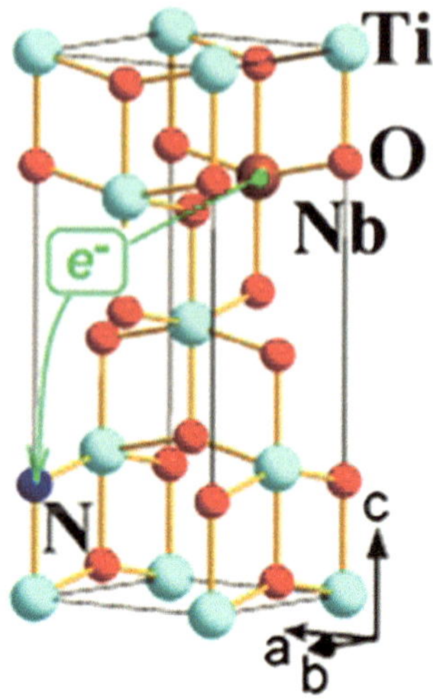

Fig. 7 Representing substitutional Nb providing an electron to substitutional N. Reprinted with permission from Ref. [30]. Copyright (2014) American Chemical Society

The co-doping of titania by N–Nb species has been recently discussed in a theoretical study with respect to its photo-electrochemical properties [33].

On the experimental side, Breault and Bartlett [42, 43] discussed heavily doped TiO_2 systems. Keller et al. [44] prepared N–Nb codoped titania nanotube arrays which show relevant visible light photo-electrochemical activity due to contents up to 15 % of Nb in the near-surface region. Lim et al. [45] carried out a work relative to a conventionally N–Nb codoped TiO_2 system, finding synergistic photocatalytic effects both for oxidation and for reduction of water pollutants when using SG-prepared samples with $0.001 < \text{Nb/Ti}$ molar ratio <0.03.

A possible electronic mechanism has been proposed [30] to explain this evidence.

Indeed, it has been reported that substitutional Nb in anatase can transfer an electron to the low-energy valence state of the co-dopant N, located near the valence band (Fig. 7).

This intrinsic charge compensation mechanism was substantiated by EPR, which showed a reduced EPR spectral intensity of $N_b^{\bullet}$ (labeled by * in Fig. 8) in the presence of Nb co-dopant.

However, we must outline that no quantitative evaluation of the amount of $N_b^{\bullet}$ is allowed by these EPR results. On the other hand, previous studies have shown that the actual nitrogen content in TiO_2 lattice is generally much lower than the nominal amount [26, 46]. Therefore, nitrogen bulk quantification remains a quite challenging task [31]. Bartlett and coauthors [42] reported the absence of any Ti^{3+} in XPS measurements, whereas the C to E lines of Fig. 8 were attributed to this species. The last attribution is also supported by the fact that these features cannot be detected at room temperature. Indeed, De Trizio et al. [47] reported that Ti^{3+} signals cannot be observed above 100 K due to the occurrence of carrier detrapping.

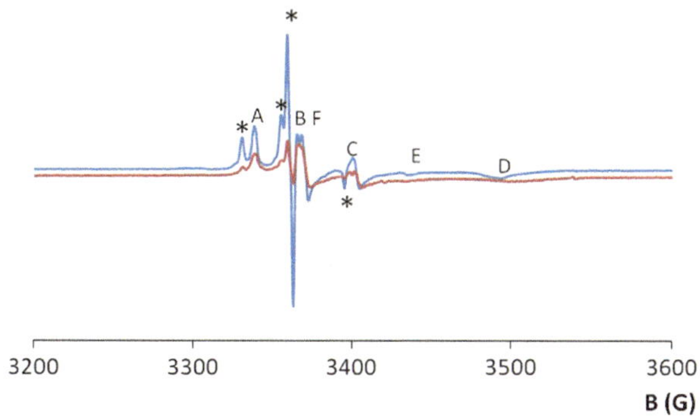

Fig. 8 EPR spectra collected at 77 K. (*) $N_b^{\bullet}$ ($g_x = 2.066$, $g_y = 2.0054$, $g_z = 2.004$, $A_x \cong A_y \cong 0$, $A_z \cong 32.2$ G); *A*, *B* ($g_x = 2.0082$, $g_y = 2.0118$, $g_z = 2.0285$) might be related to paramagnetic radical oxygen species, while the lines *C* ($g_1 = 1.98$), *D* ($g_2 = 1.93$), and *E* ($g_3 = 1.96$) can be attributed to Ti^{3+} in different environments. The F line could be attributed to F^+ color centers. *A* to *F* lines are not detectable at room temperature. Adapted with permission from Ref. [30]. Copyright (2014) American Chemical Society

However, these features decrease or completely disappear in all the N–Nb codoped samples, supporting the occurrence of charge compensation between Nb (acting as electron donor), involving also $N_b^{\bullet}$ (which would become diamagnetic N_b^-), as predicted by calculations [30, 33].

At last, only the A and B lines due to oxygen radical and the C to E lines due to Ti^{3+} species were appreciable in the Nb-doped sample at 77 K (private communication), in agreement with the presence of Ti^{3+} species in the bulk of these systems reported elsewhere [30, 47], as well as with calculations [30, 33].

2.2 *Surface Oxygen Species Adsorbed on K-Doped $SrTiO_{3\pm\delta}$*

Transition metal oxide mixtures with perovskite-like structure are well known as catalysts for the flameless combustion (CFC) of methane. Thermal stability and high surface area of these materials can be achieved by preparing them as nanosized particles, calcined at very high temperature [54]. Nanosized $SrTi0_{3\pm\delta}$ particles prepared by flame hydrolysis (FH) and partially substituted with potassium (KFH) or with gadolinium (GFH) for strontium have been investigated [55, 56]. A different ionic charge of potassium (1+) and gadolinium (3+) with respect to strontium (2+) was inducing a nonstoichiometric oxygen composition, which could affect the catalytic activity. Only the former case will be taken into consideration in this paragraph.

The X-band EPR spectrum of the $Sr_{0.9}K_{0.1}TiO_{3\pm\delta}$ (KFH) sample was detected both with X-band and with Q-band EPR. It was composed of a very narrow Lorentzian-shaped line at $g \cong 2.008$, with a temperature-independent peak-to-peak width ΔH_{pp} of 1.6 G (X-band) and of 3.4 G (Q-band) (Fig. 9).

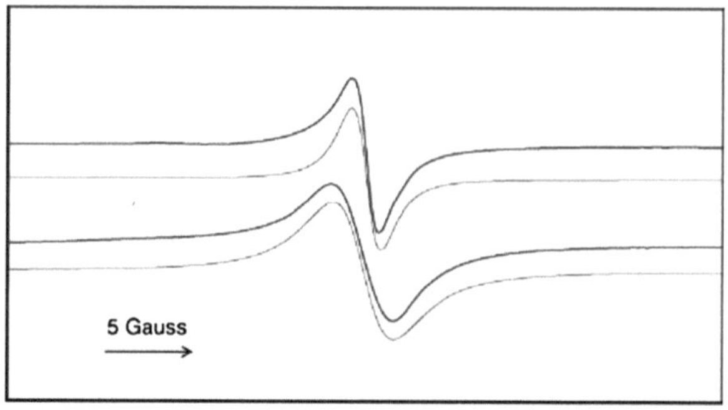

Fig. 9 X-band (*upper*) and Q-band (*lower*) EPR spectra of KFH sample, recorded at 290 K. The thinner lines are Lorentzian-shaped simulations. Reprinted with permission from Ref. [55]

The partial substitution of K^+ for Sr^{2+} could be counterbalanced by a partial oxidation of Ti^{3+} to Ti^{4+}. Indeed, it was the case with the $Sr_{0.9}K_{0.1}TiO_{3\pm\delta}$ (KFH) sample. The single line at $g \cong 2.008$ was then attributed to Ti^{4+}/O_3^- in which a Ti^{4+} ion would interact with an ozonide radical, as already reported in analogous literature cases [57–63]. Surface O_3^- is characterized by high mobility ("suprafacial" α oxygen) showing a "motional narrowed" Lorentzian-shaped spectrum. An exchange frequency exceeding 17 MHz, provided by the thermal energy of *ca.* 10^{-21} J at 100 K, was evaluated in this case to mix the spectral anisotropies and produce a Lorentzian-shaped line.

Furthermore, this line was disappearing after catalytic use further supporting its attribution to surface species involved in the catalytic reaction, whereas the bulk species formed when Gd^{3+} ion was doping $SrTiO_{3\pm\delta}$ (see below) seemed not much involved in the CFC of methane, even causing a worsening of the catalytic activity.

By concluding, it was reported that the partial substitution of K^+ for Sr^{2+} in $SrTiO_3$ caused an improvement of the catalytic activity, though only at $T < 550$ °C. Indeed, the oxidation of Ti^{3+} to Ti^{4+}, forced by the partial substitution with the single valence K^+ ion, was favoring the formation of the oxidant species Ti^{4+}/O_3^-. The O_3^- ions were almost free to move on the KFH surface ("supra-facial" oxygen species) as indicated by their narrowed Lorentzian-shaped EPR line.

2.3 *The Strange Case of Doped Nanoparticles of Cuprates*

Cu^{2+} is a regular component of $La_{1-x}M_xCuO_4$ (M-doped) cuprates. Therefore, it is surprising that in many cases these systems show the EPR spectrum typical of diluted Cu^{2+}, i.e., with a resolved Zeeman and hyperfine structure, instead of displaying a single EMR band due to spin–spin coupling among adjacent Cu^{2+} ions (as shown later in the next paragraph for few cases of doped cuprates), or even showing no-EPR spectrum at all, as expected for antiferromagnetic systems, due to adjacent Cu^{2+} ions interacting with each other by superexchange through O bridges. EPR patterns like that above shown in Fig. 1 would arise only from Cu^{2+} ions which are

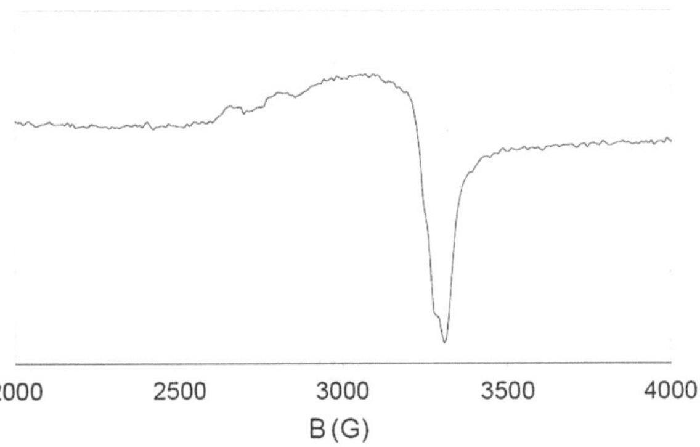

Fig. 10 The EPR spectrum of fresh $La_{1.8}Sm_{0.2}CuO_4$ detected at 120 K. $g_{//} = 2.35$; $g_{\perp} = 2.06$; $A_{//} = 144$ G. Reprinted with permission from Ref. [50]

isolated. Indeed, this was the case of the EPR spectrum which was accompanying the formation of solid solutions of La_2O_3 in CeO_2 (molar ratio 1:2) to which CuO was added (La:Cu ratio 10:1) [48] An analogous pattern was detected also with $(CeO_2)_{1-y}(La_2CuO_4)_y$ ($y = 0.25$; 0.43) [49]. In both those cases it has been demonstrated that the EPR spectrum was coming from Cu^{2+} ions inserted into the crystal framework.

A similar situation was observed also with 20–40 nm nanoparticles of $La_{1.8}M_{0.2}CuO_4$ (M = Pr^{3+}, Sm^{3+}, or Tb^{3+} [50] (Fig. 10).

Unfortunately, in the last case the Cu^{2+} EPR spectra were characterized by poor intensity and resolution, so that they were not able to report any significant difference between distortions occurring in the Cu^{2+} neighbors because of the different ionic radii of the Pr^{3+}, Sm^{3+}, or Tb^{3+} substituting for La^{3+}.

Pr^{3+}, Sm^{3+}, and Tb^{3+} are paramagnetic ions [51]. However, their spin–lattice relaxation rate is generally so high that their contribution to the EPR pattern is not expected at temperature higher than 20 K. Indeed, only lanthanide ions in the S-state (L = 0; $4f^7$), i.e., Eu^{2+}, Gd^{3+}, and Tb^{4+}, can show an EPR spectrum at higher temperature [52, 53]. However, no Tb^{4+} was noticed in the $La_{1.8}Tb_{0.2}CuO_4$, indicating that Tb^{3+} only was present in it.

A broad band was added to the EPR spectrum shown in Fig. 10. This pattern was not attributable to isolated paramagnetic ions but, instead, to the formation of ferromagnetic domains. Therefore, it will be discussed in the next paragraph.

3 Nanosized-Doped Particle Systems and Collective Phenomena

3.1 Ferromagnetic Domains of Sm^{2+} Ions in Nanosized Sm-Doped Particles

When analyzing the EPR spectrum of $La_{1.8}Sm_{0.2}CuO_4$ in a magnetic field range wider than that of Fig. 10, a new very intense feature adds to that there shown. This

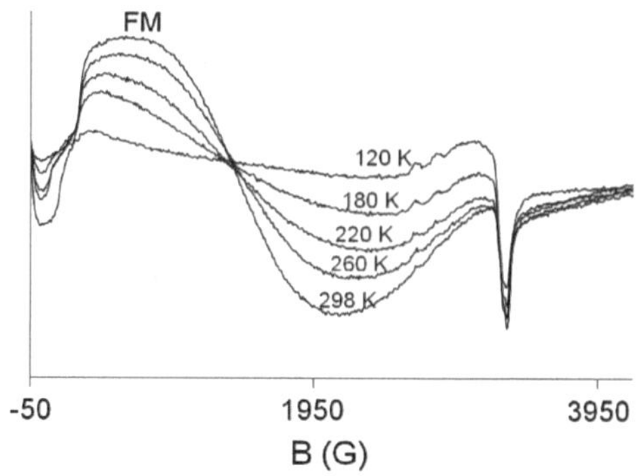

Fig. 11 The temperature-dependent FM band observed with $La_{1.8}Sm_{0.2}CuO_4$ attributed to ferromagnetic resonance. Adapted with permission from Ref. [50]

broad band, labeled by FM in Fig. 11, is not attributable to Sm^{3+} ions. Furthermore, it has been reported [50] that it undergoes hysteresis effect, so that it is not attributable to any EPR resonance at all but, instead, to ferromagnetic resonance (FMR) [64].

This means that FM domains form in this sample, generating an internal field $B_{\rm i}$ and, therefore, a spectral shift toward lower fields [65]. Such systems should not be distributed uniformly throughout the bulk, as suggested by the dependence of their spectrum on sample orientation. FM clusters have been reported also with other Sm-containing perovskites, namely with $Sm_{0.2}Ca_{0.8}Mn_{1-x}Ru_xO_3$ [66], and attributed to the formation of "Bound Magnetic Polarons" (BMP), also indicated as "Ferrons" [67]. These systems should form in the presence of electrons concentrated enough to form local FM "droplets," growing up in the vicinity of electron donors, like Sm^{2+} ions. The last would be created through reactions like

$$10\,La_{1.8}Sm_{0.2}CuO_4 \rightarrow 9\,La_2O_3 + 10\,CuO + Sm_2O_3 \tag{4}$$

$$Sm_2^{3+}O_3^{2-} \rightarrow 2\,Sm^{2+}O^{2-} + \frac{1}{2}\,O_2 \tag{5}$$

Sm^{2+} are diamagnetic ($J = 0$) $4f^6$ ions in their fundamental state. However, the energy necessary to excite one of their electrons so to obtain a $J = 1$ state is less than 300 cm^{-1}, i.e., comparable with the room temperature energy ($\sim$200 cm^{-1}). This value is very low if compared to that required by the other rare-earth ions, which generally amounts to some thousands cm^{-1}. This facility to obtain parallel electron spins with Sm^{2+} ions can account for the formation of ferromagnetic systems when some ions of this kind cluster together.

The FM bands were no more detected after sample calcination at 800 °C. On the other hand, XRD patterns revealed that La_2O_3 was no more present in these cuprate samples after such a treatment. This was in line with the above-reported attribution of the FM band to systems forming only when reactions (4) and (5) occur.

3.2 *Superparamagnetic Nanoparticles with Fe-Doped Zeolites and Silicalites*

3.2.1 Isolated Fe^{3+} Ions and Their Clusters in Fe-Doped Zeolites and Silicalites

Iron-containing zeolites with MFI structure have attracted considerable interest since 1988, when three groups of researchers [68–70] independently found that these materials are effective catalysts in the one-step hydroxylation of benzene to phenol by N_2O, a process much less expensive and less troublesome than some previously adopted ones [71]. Fe(III) is a fundamental constituent of the catalyst [71–74], so that these systems have been widely investigated by EPR spectroscopy [72, 73, 75–93].

Their EPR spectrum is due to the high-spin (S = 5/2) Fe(III) ions. It can be described by the Hamiltonian

$$\hat{H} = g\beta \mathbf{S} \cdot \mathbf{B}_0 + D\left[S_z^2 - \frac{1}{3}S(S+1)\right] + E\left(S_x^2 - S_y^2\right) \tag{6}$$

where $g \cong g_e$ since high-spin Fe(III) is a ^6S ion, $\mathbf{B}_0$ is the static magnetic field vector, $D = 3\ D_z/2$, $E = (D_x - D_y)/2$, and D_i ($i = x, y, z$) are the principal values of the zero-field splitting (ZFS) interaction, which are ordered so that $|D| \geq |E/3|$ and $D{\cdot}E \leq 0$ [94]. For future convenience we also define the dimensionless parameter $\lambda = 3\ |E/D|$, $0 \leq \lambda \leq 1$.

The ZFS interactions for a high-spin Fe^{3+} d^5 ion are sensitive to its environment because of their dependence on the ligand field [95]. The X-band EPR spectrum of iron-doped zeolites and silicalites usually comprises several lines. A group of them is observed at low resonance field, i.e., at $g_{\text{eff}} \geq 4$ with a prominent peak at $g_{\text{eff}} = 4.3$ and several other ones down to $g_{\text{eff}} \cong 9$. Two major broad lines are found at $g_{\text{eff}} \cong 2.0$ and 2.3, the two peaks being largely superimposed. Other minor features are observed, including a peak very close to zero field ($g_{\text{eff}} \geq 10$). Spectra of this kind obtained with Fe–silicalite crystals of 0.5–1.0 μm [96] are reported in Fig. 12.

The conventional attribution of the lines of Fig. 12 was as follows [73, 75, 78, 80, 84–86, 88, 90–92]: the $g_{\text{eff}} \geq 4$ group was attributed to isolated Fe(III) ions in the zeolitic framework, undergoing various degrees of distortion and the $g_{\text{eff}} \cong 2.0$ and 2.3 lines were attributed to oxidic iron clusters and particles where Fe(III) ions were strongly interacting with each other (earlier interpretations [73, 75, 77] attributed the $g_{\text{eff}} \cong 2.0$ to extra-framework ions in octahedral symmetry). However, this interpretation often was not in agreement with data from other techniques [79].

Therefore, it was later assumed [76, 77, 79, 89] that both framework and extra-framework Fe(III), either as isolated ions or as aggregates (oligo-nuclear clusters, oxidic particles), were contributing to each of the above spectral features. Indeed, the ZFS interactions of Fe(III) ions in MFI zeolitic catalysts are largely inhomogeneous in nature. Therefore, good spectral simulations were obtained for

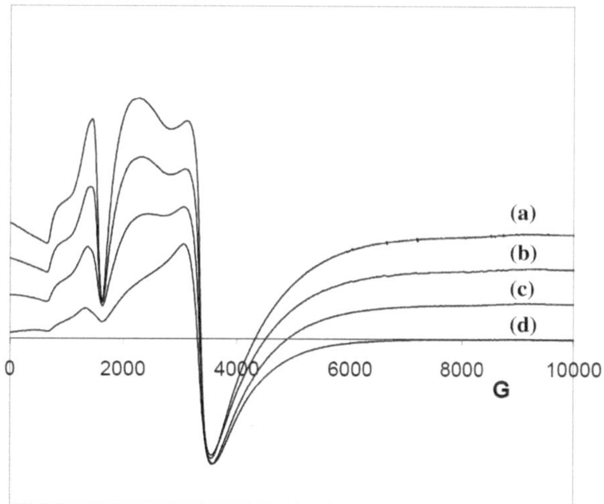

Fig. 12 EPR spectra of Fe-doped silicalite, detected at 110, 150, 200, 295 K ((*a*)–(*d*)). Adapted with permission from Ref. [96]

the EPR spectra of Fe(III)-doped MFI Zeolites by adopting a *distribution* of the ZFS interaction parameters [93, 97, 98]. Many different intra- and extra-lattice sites are possible for Fe(III) ions. The latter include both sites within the channels and on the external surface of the crystallites, either as isolated ions or as oxidic aggregates.

Aiming at a better evaluation of the contributions of Fe(III) ions in extra-lattice sites, these samples underwent a steaming process, i.e., a treatment in a flow of 75 mol% steam and 25 mol% nitrogen at 773 K for 6 h. Indeed, after this treatment of the Fe-doped silicalite and of the MFI Zeolite samples, new FMR features were observed at magnetic field values lower than ca. 2000 G and attributed to oxidic aggregates [96, 99] (see, for the former sample, Fig. 13). These lines were further affected by the following calcination process (see Fig. 14).

The $g \cong 2$ line (detected at a magnetic field of ca. 3000 G) will be discussed below.

3.2.2 Superparamagnetism of Nanosized Fe-Containing Clusters

The above-mentioned steaming process was setting a large fraction of iron into extra-lattice sites. Those ions were organizing into nanometric superparamagnetic particles. Indeed, the $g \cong 2$ signal of Figs. 13 and 14 was approximately Lorentzian at room temperature, but it became broader and more asymmetric with decreasing temperature. This behavior is typical of single-domain superparamagnetic particles [100], as reported for silicate glasses containing nanosized clusters of magnetite [101], for γ-Fe_2O_3 [102], for iron-doped borate glasses [103], and for nanosized perovskites [104]. Superparamagnetic particles are single-domain nanosized magnetic systems. They form when the single-domain state is preferred to a multi-domain one, because the energy required to the formation of the former is

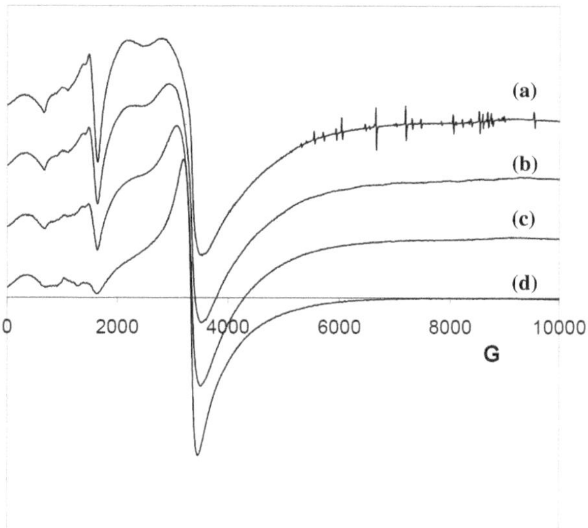

Fig. 13 EPR spectra of Fe-doped silicalite after steaming. Detection temperature 110, 150, 200, 295 K ((*a*)–(*d*)). Narrow lines in (*a*) indicate the presence of physisorbed molecular O_2. Adapted with permission from Ref. [96]

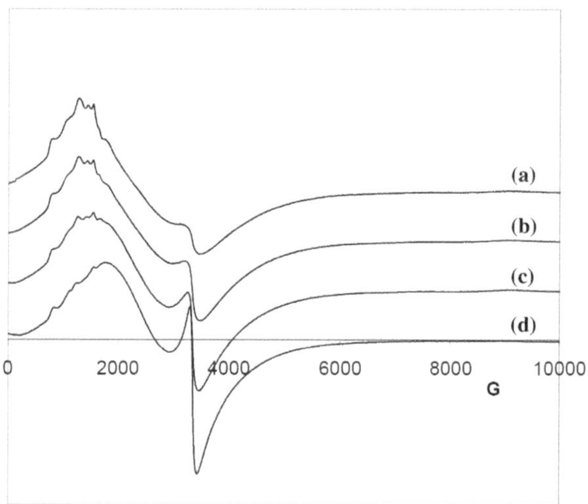

Fig. 14 EPR spectra of Fe-doped silicalite after sample steaming followed by recalcining. Detection temperature 110, 150, 200, 295 K ((*a*)–(*d*)). Adapted with permission from Ref. [96]

smaller than that which would be needed to form the latter with its domain walls. Such particles can exhibit superparamagnetic behavior above a certain critical temperature, below which they are in the normal (anti)ferromagnetic state [100]. The resonance condition for an isotropic superparamagnet is

$$\omega_c = \gamma \boldsymbol{B} \tag{7}$$

where γ is the magnetogyric ratio and $\omega_c \equiv 2\pi\nu_c$. On the other hand, for anisotropic superparamagnetic particles, assuming that the anisotropy energy is smaller than the coupling to the external field, the resonance condition is

$$\omega_c = \gamma[B + B_a P_2[cos(\psi)]L_2(\xi)/L(\xi)] \quad (8)$$

where B_a is the anisotropic magnetic field, ψ is the angle between the anisotropy axis and the external field, P_2 is the second degree Legendre polynomial, $\xi = M_{nr}VB_a/kT$, M_{nr} is the non-relaxing magnetization, V is the particle volume, $L(\xi)$ is the Langevin function, and $L_2(\xi) = 1 - 3\ L(\xi)$. Equation (8) shows that thermal fluctuations lead to a decrease of the effective anisotropy field (in addition to any actual temperature dependence of the anisotropy constant), which can be written as

$$B_{a,\text{eff}} = B_a L_2(\xi)/L(\xi) \quad (9)$$

It follows, from asymptotic behavior for $\xi < 1$, that $B_{a,\text{eff}}$ decreases with increasing temperature, being $B_{a,\text{eff}} \propto \xi \propto 1/T$. Therefore, from Eq. (8) the effect of increasing temperature on EPR spectra is that the resonance field approaches the isotropic value $\omega_c = \gamma B$. For an actual fine-particle system, with distribution of particle volume and easy-axis orientation, the dependence of the resonance condition upon the direction of the anisotropy axes results in an asymmetric lineshape at low temperature, accompanied by a shift of the resonant field. With increasing temperature, the linewidth reduces and the line tends to a Lorentzian shape with isotropic resonance field $B_r = \omega_c/\gamma$, as thermal fluctuations average to zero the anisotropy field. With decreasing temperature, the linewidth increases and the shape becomes more and more distorted, whereas the narrow line tends to vanish.

The narrow superparamagnetic line at $g \cong 2$ is exhibited by those particles for which $\tau_{sp} < \tau_L$. τ_{sp} is the superparamagnetic relaxation time, due to the thermal fluctuations of the direction of magnetic moment in single-domain particles smaller than a "critical size." τ_L is the Larmor precession time in the magnetic resonance field H_o. For these very small particles, when the anisotropy energy is much smaller than kT, the following relations hold [105]:

$$\tau_{sp} \cong M_S V/(\gamma_0 k_B T) \qquad \gamma_0 = 2\pi/(\tau_L H_0) \quad (10)$$

where M_S is the saturation magnetisation (expressed in A m^{-1}), V is the volume of the particle undergoing the fast relaxation process, γ_0 is the gyromagnetic ratio, and k_B is the Boltzmann's constant. The magnetic anisotropies are "motionally" averaged out at temperature high enough to make $\tau_{sp} < \tau_L$. In this case, the EPR line narrows by the temperature-dependent factor:

$$f \cong \tau_{sp}/\tau_L \cong M_s V\, H_0/(2\pi\, k_B\, T) \quad (11)$$

In real systems a distribution of particles volume occurs around a mean value $\langle V \rangle$, the particle diameter ranging between that of a single-domain particle (<15–20 nm) and that of a superparamagnetic assembly of atoms (≅1 nm). A co-existence of ferrimagnetic and superparamagnetic systems is expected [106], the former causing a broader EPR feature, and the latter a motionally narrowed one. The equilibrium between these two species depends also on the sample preparation method [107]. The peak-to-peak linewidth ΔH_{pp} of the narrowed line decreases at high temperature, following Eq. (11). Therefore, in the regime of "motional" narrowing, the EPR linewidth should be expressed by

$$\Delta H = \Delta H_o + \Delta H_o \gamma V/T \quad (12)$$

where ΔH_o is the linewidth for $T \to \infty$. In Eq. (12) $\Delta H = H_{x,y} - H_z$, where $H_i = h\nu/g_i\mu_B$ ($i = x$, y or z), h is the Planck's constant, ν is the instrument frequency, and μ_B is the Bohr's magneton and

$$\gamma = M_S H_o/(2\pi \mu_o k_B) \quad (13)$$

where M_S is the saturation magnetisation value, H_o is the resonance magnetic field (in tesla) and $\mu_o = 4\pi\ 10^{-7}$ (weber $A^{-1}m^{-1}$) is the vacuum permeability. μ_o must be introduced if M_S is expressed in tesla (as usual) instead of in ($A\ m^{-1}$), as it should be.

For example, a value $M_S = 0.048$ T is reported in literature [108] for Fe_3O_4. Equation (12) can be compared with the experimentally determined straight line

$$\Delta H = \alpha + \beta/T \quad (14)$$

leading to

$$V = \beta/(\alpha\gamma) \quad (15)$$

and hence to a qualitative evaluation of the particle diameter d.

In the above-mentioned examples, extra-framework nanosized superparamagnetic particles were forming upon micro-sized Fe-doped zeolites.

Extra-framework superparamagnetic particles of similar kind form also with *ca.* 20-nm nanosized Fe-doped silicalites [109]. An even more detailed analysis was carried out in that case on the nature of these superparamagnetic particles, by comparing EPR to temperature-programmed reduction (TPR) results. After sample steaming, two different kinds of extra-framework Fe^{3+} ions were singled out by the latter technique. One of them (reducing to Fe^{2+} at 643 K) was attributed Fe_2O_3 entities, whereas the latter (reducing to Fe^{2+} at 823 K) was attributed to charge-compensating Fe^{3+} ions.

The EPR spectra were collected before and after reducing the sample in a 40 cm^3/min stream of 10 % H_2 in He, while increasing temperature by 10 K/min up to either 673 or 873 K, then maintained for 5 min, before cooling down to room temperature in pure flowing He.

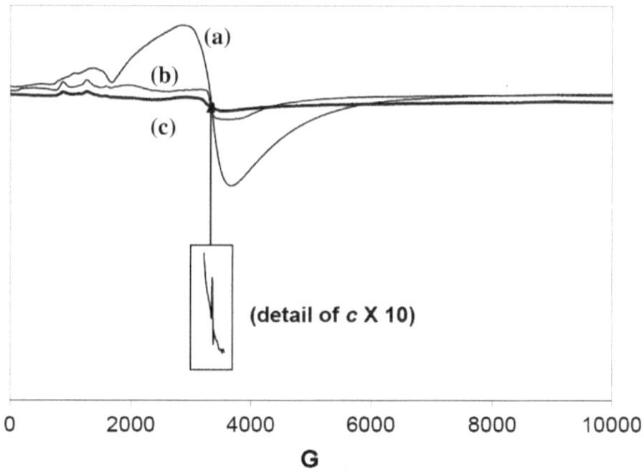

Fig. 15 EPR spectra of Fe-doped silicalite after steaming: **a** as prepared, **b** after reduction up to 673 K, **c** after reduction up to 873 K. (The *inset* shows an EPR line attributed to traces of organic radicals.) Reprinted with permission from Ref. [109]

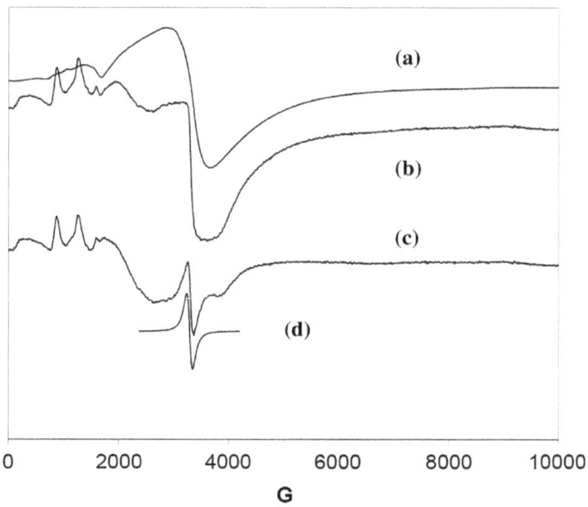

Fig. 16 EPR spectra of Fe-doped silicalite after steaming: **a** as prepared, as in Fig. 15a, b after reduction up to 673 K as in Fig. 15b, but reported to the same intensity of (**a**). **c** Difference between (**b**) and (**a**). **d** Computer simulation of spectrum in 16 (**c**), with g = 2.04, peak-to-peak linewidth =110 G, Lorentzian shape. Reprinted with permission from Ref. [109]

The spectrum of the original, unreduced, sample was formed, as expected, of two main features, at $g > 4$ and at $g \cong 2$, respectively (Fig. 15a). After reduction of the catalyst up to 673 K, the intensity of the whole pattern decreased by *ca.* 80 % and a better resolution was obtained in the low-field region, which resulted composed of two main lines, at $g = 8.2$ and 5.56, respectively (Fig. 15b). The spectral intensity further decreased (by *ca.* 92 %, with respect to the unreduced sample) after reduction of the sample up to 873 K (Fig. 15c).

Then, the authors compared the shape of the spectral profile before and after TPR, by renormalizing the spectra of Fig. 15a, b, as shown in Fig. 16a, b, and subtracting the former from the latter. The spectrum of Fig. 16c was so obtained, which resulted a 110 G broad Lorentzian line, centered at $g \cong 2$ (Fig. 16d). After reduction up to 873 K this Lorentzian-shaped line disappeared, indicating that it was due to species no more present at T > 823 K. Therefore, it was possible to conclude that the superparamagnetic particles were due to the extra-framework Fe_2O_3 entities in which the Fe^{3+} ions were reducing to Fe^{2+} at 643 K, as established by TPR.

3.3 *Gd^{3+} Nanosized Ferromagnetic and Superparamagnetic Clusters in Gd-Doped $SrTiO_{3\pm\delta}$*

Nanosized $SrTi0_{3\pm\delta}$ particles prepared by flame hydrolysis (FH) and partially substituted with gadolinium for strontium (GFH samples) have been investigated [54, 55]. The different ionic charges of gadolinium (3+) with respect to strontium (2+) were inducing a nonstoichiometric composition, which could affect the catalytic activity.

The X-band EPR spectrum of the (GFH) $Sr_{0.9}Gd_{0.1}TiO_{3\pm\delta}$ sample showed a high-intensity, nearly symmetric band at room temperature, typical of ferromagnetic (FM) resonance of Gd^{3+} ion domains (Fig. 17) which was about 100 times broader than the EPR narrow line observed with K^+-doped samples (see paragraph. 1.2).

With decreasing temperature, this pattern broadened and its intensity increased, though less than how expected on the base of the Curie's law, indicating that the single ions are indeed reciprocally interacted in a ferromagnetic way. Furthermore, at lower temperature the presence of bumps like B on the left part of the spectrum was becoming more evident (Fig. 18), indicating the presence of internal magnetic fields in FM systems with positive axial anisotropy [110].

The EPR spectrum of this sample was even more symmetric when analyzed at the Q microwave band and, in that case, it was almost perfectly fitting the sum of two Lorentzian lines at g = 1.991 and 2.020, width of 604 G and of 1600 G, respectively, broadening with decreasing temperature and with integral area ratio of

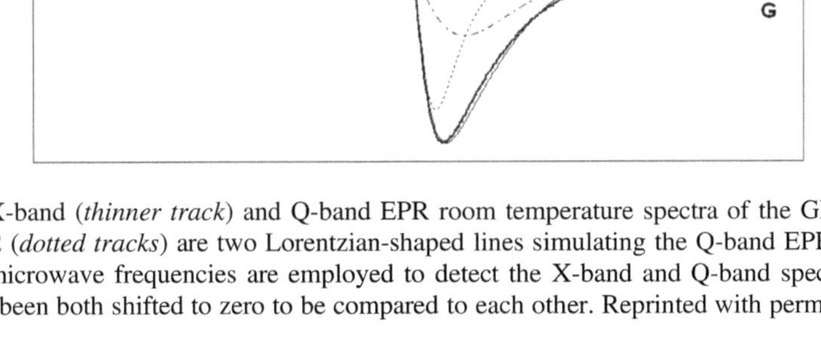

Fig. 17 X-band (*thinner track*) and Q-band EPR room temperature spectra of the GFH sample. L1 and L2 (*dotted tracks*) are two Lorentzian-shaped lines simulating the Q-band EPR line. Two different microwave frequencies are employed to detect the X-band and Q-band spectra, so that they have been both shifted to zero to be compared to each other. Reprinted with permission from Ref. [55]

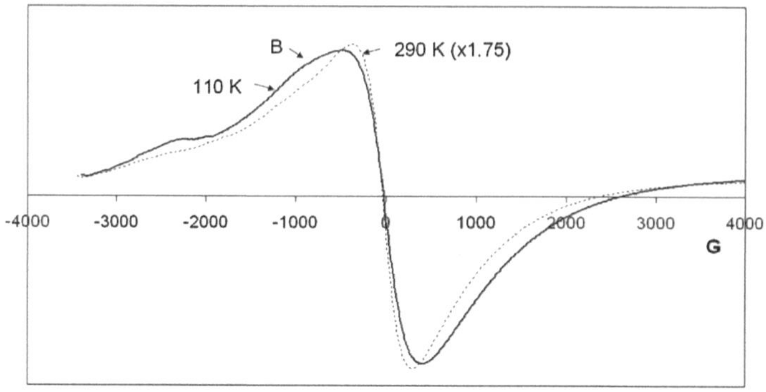

Fig. 18 X-band EPR spectra of GFH sample, recorded at 140 K (*thicker track*) and 290 K (*dotted track*, multiplication factor 1.75). Reprinted with permission from Ref. [55]

4:1 at room temperature (Fig. 17) and of 1.5:1 at 100 K. The presence of bumps like B in the X-band spectrum, and not in the Q-band one, was further confirming their attribution to FM systems with magnetic anisotropy $\Delta H_a \equiv |H_{//} - H_{\perp}|$. Indeed, the following equation holds [110]:

$$\Delta H_{\mathrm{a}} = \alpha'\{[\nu_{\mathrm{Q}}\Delta H_{\mathrm{pp}}(\mathrm{X}) - \nu_{\mathrm{X}}\Delta H_{\mathrm{pp}}(\mathrm{Q})]/(\nu_{\mathrm{Q}} - \nu_{\mathrm{X}})\} \tag{16}$$

where $\alpha' = 1.732$ for Lorentzian shape. Since $\Delta H_a \geq 0$, Eq. (16) requires $\Delta H_{\mathrm{pp}}(\mathrm{X}) \geq 1/3\ \Delta H_{\mathrm{pp}}(\mathrm{Q})$, as in this case, where $\Delta H_{\mathrm{pp}}(\mathrm{X}) > \Delta H_{\mathrm{pp}}(\mathrm{Q})$. However, no reliable quantitative evaluation of the magnetic anisotropy ΔH_a(GFH) can be obtained, due to the low spectral resolution (Fig. 17).

Different preparation methods were producing particles of different sizes, so markedly affecting their EPR spectra. Indeed, the above-mentioned GFH sample (prepared by FH procedure) was 56 nm in diameter, whereas a GSG sample with the same chemical composition but prepared by sol–gel procedure was by far smaller having a mean diameter of 24 nm only.

As a consequence, both X-band and Q-band EPR spectra of the latter sample were composed, at room temperature, of a single nearly symmetric feature at $g = 1.991$, with a peak-to-peak width of *ca.* 40 G (Fig. 19), i.e., less than 1/10 that observed with the former (Fig. 17).

Furthermore, with decreasing temperature the intensity of these GSG spectra decreased a bit, instead of increasing as expected on the base of the Curie's law, while a second broader overlapping feature was appearing both in the Q-band (Fig. 20) and in the X-band spectra.

This behavior was in accordance with the particle nanometric size of the (GSG) $Sr_{0.9}Gd_{0.1}TiO_{3\pm\delta}$ sample. In fact, it has been reported [110] that with small (single-domain) particles (as in this case) at high temperature, when kT overcomes the small anisotropy energy of these particles, the FM resonance bands narrow and merge into a Lorentzian-shaped (Superparamagnetic, SPM) line like those shown in Fig. 19 and similar to the symmetric narrow feature at RT shown in Fig. 20. At

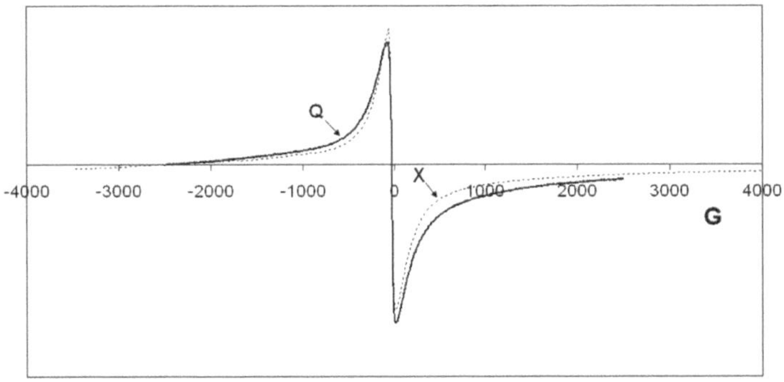

Fig. 19 Room temperature Q-band (*thicker track*) and X-band (*dotted track*) EPR spectra of GSG sample. Reprinted with permission from Ref. [55]

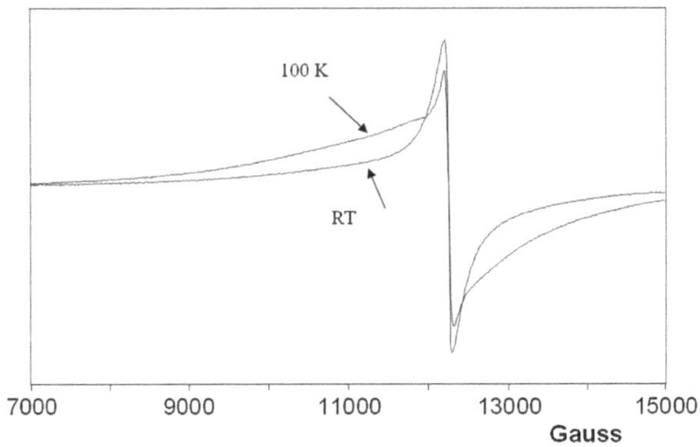

Fig. 20 Q-band EPR spectra of GSG sample, recorded at two different temperatures. Reprinted with permission from Ref. [55]

intermediate temperatures an intermediate regime must occur in these systems, in which a "two-line pattern" as that of GSG at 100 K (Fig. 20) appears, formed of a broad feature overlapping a narrower one [111–114]. The relative intensity of these two spectral contributions depends on particle size and shape distribution function, as well as on the magnitude of the magnetic anisotropy.

3.4 Polarons and Polaritons Propagation with Nanosized Particles of Gd-Doped CeO_2

Ce^{4+} ions in CeO_2 are diamagnetic, so that no-EPR spectrum is expected from this material, when undoped. In principle, oxygen-deficient CeO_2 could contain some Ce^{3+} ions. However, the last has an unpaired 4f electron with a short spin–lattice relaxation time [115–121] so that its EPR spectrum could be expected at very low temperature only. By contrast, nano-structured fluoritic $Ce_{1-x}Gd_xO_{(4-x)/2}$ have been intensively studied, also by EPR, as solid electrolytes for solid oxide fuel cells

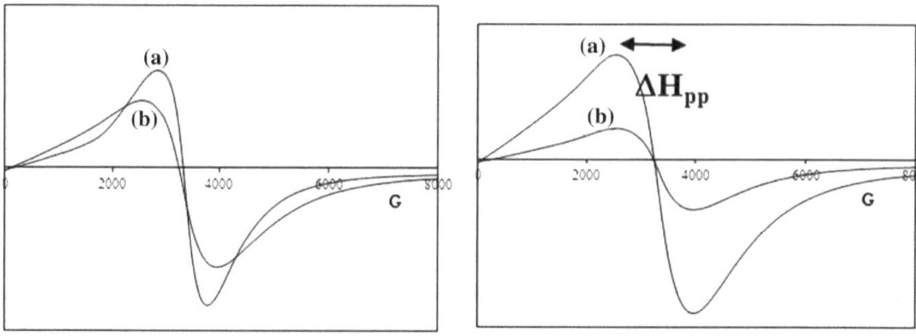

Fig. 21 EPR spectra of "as prepared" $Ce_{1-x}Gd_xO_{(4-x)/2}$. *Left*, T = 110 K ((*a*) $x = 0.10$; (*b*) $x = 0.20$): keeping constant T, *broader lines* are detected at higher x values. *Right*, $x = 0.20$ ((*a*) T = 110 K; (*b*) T = 380 K): keeping constant x, more *intense lines* are detected at lower T values. Adapted with permission from Ref. [126]

(SOFCs) [122], being ion conductors even better and able to operate at lower (500–700 °C) temperatures [123] than conventional yttria-stabilized zirconia. The investigated nanosized $Ce_{1-x}Gd_xO_{(4-x)/2}$ samples were characterized by a very high specific surface area, so that their thermodynamics and defect equilibria were quite different from that of bulk materials [124, 125].

Some selected EPR spectra of these systems [126] are shown in Fig. 21. All of them were composed of a single line at $g \cong 2.02$. In any case, the spectral shape was only approximately Lorentzian, because their left part was always a bit broader than the right one. No successful simulation was obtained for them, though the bump on their left side could have the same ferromagnetic origin as that observed with a $Sr_{0.9}Gd_{0.1}TiO_{3\pm\delta}$ sample above discussed in the section devoted to Gd-doped titanates (see Refs. [55, 127]).

At $T = 110$ K the spectral intensity I was 1.5 times larger with the $x = 0.10$ than with the $x = 0.20$ sample (Fig. 21, *left*). Such a different spectral intensity was connected to a different ΔH_{pp} peak-to-peak linewidth which was broader at higher x values, due to a shorter spin–spin relaxation time T_2 caused by the stronger interaction occurring among more concentrated Gd^{3+} ions. By contrast, the EPR linewidth and shape were independent on the temperature (Fig. 21, right), excluding the presence of temperature-dependent phenomena affecting T_2.

The Curie's law would hold for a system of non-interacting paramagnetic ions. Therefore, only in that case the magnetic susceptivity χ (and so the spectral intensity $I(T)$) would vary with temperature following the trend:

$$I(T) \cdot T = I_{110\ \mathrm{K}} \tag{17}$$

where $I_{110\ K}$ indicates a spectral intensity kept as reference, which in [68] was that measured at the lowest (110 K) temperature. Deviations from the Curie law could be described by

$$I(T) \cdot T = I_{110\ K} \cdot T^{\alpha} \tag{18}$$

Values of α greater than 0 were attributed to magnetic interactions among paramagnetic ions. In this case, $I(T) \cdot T$ would increase with T [128].

A decrease of $I(T) \cdot T$ versus T (i.e., $\alpha < 0$) was attributed to thermally activated electronic conductivity σ due to small polarons, hopping between ions in different valence states [129, 130], i.e., to conduction electrons propagating in a *percolative* way between localized states randomly distributed in energy and position [131]. In fact, any electronic contribution to conductivity would cause a *skin depth* on the surface of a sample undergoing a microwave magnetic field, as when detecting an EPR spectrum. In a *percolative* regime, σ increases with increasing temperature. Therefore, a thinner *skin depth* would arise at higher temperature, causing an EPR signal intensity lower than expected on the basis of the Curie law only, so that the product $I(T) \cdot T$ would decrease with increasing T, instead of being constant as with Eq. (17). This is the meaning of the a $\alpha < 0$ value measured instead of $\alpha = 0$ observed with the "as prepared" $x = 0.10$ sample, and perhaps also after its oxidation at 473 K.

This *percolative* conduction regime can be accounted for by the local structural disorder introduced by the Gadolinium doping [132, 133] and by the oxygen vacancies created by the gas–solid equilibrium occurring with the sample. This deviation from Curie's law reduced or disappeared at all after thermal treatment of the $x = 0.10$ sample with O_2 and the parameter α became even positive after oxidation of this sample at 773 K. At the contrary, the α parameter was always positive with the $x = 0.20$ sample, independently from any annealing procedure. All the EPR results were identical 1 year later, indicating a great stability of those systems. Therefore, the chemical formula of the samples was more properly written as $Ce_{1-x}Gd_xO_{2-x/2-\delta}$, where δ was a function of x, T, and $P(O_2)$. The electronic conductivity σ was increasing with increasing conduction electron concentration, i.e., with decreasing oxygen partial pressure and/or gadolinium x concentration. All the EPR experimental results, being related to σ, were interpreted by defect equilibria. The last, in turn, were depending not only on annealing conditions and x doping level, but also on the sample size.

All the samples investigated in [126] were *ca.* 20 nm nano-crystalline.

Crystalline samples of $Ce_{1-x}Gd_xO_{2-x/2-\delta}$ *ca.* 100 nm in size and with $0 < x < 1$ were also investigated in [134].

On the local scale, Gd- and Ce-rich nanosized *droplets* (i.e., small regions, namely some Ångstroms wide) were forming in those samples, exhibiting either a C-type (Gd_2O_3) or a distorted fluorite (CeO_2) structure in the whole compositional range. For $x \leq 0.25$, the Gd-rich *droplets* were building C-type nano-domains embedded in the distorted fluorite (CeO_2) matrix. At $x \approx 0.311$ these C-type nano-domains were percolating inside each crystallite and a structural phase transformation occurred to a C*phase (up to $x \leq 0.433$) and then to a C phase at higher x values. At $x \approx 0.311$ the peak-to-peak EPR linewidth ΔH_{pp} (always increasing with increasing T) showed a step-like broadening of *ca.* 800 G, attributed to the increased Gd–Gd dipolar interactions characterizing the new crystal structure. This topic has been further investigated by other techniques confirming this interpretation. All these results were further rationalized in the framework of a percolation-driven phase transition [135].

An interpretation of the EPR spectral profile was attempted at $x \geq 0.5$ only, i.e., when a single mechanism seemed to prevail to form it. However, the result was rather problematic. Indeed, at these x values the EPR pattern was assuming the shape of a single Dysonian line [134]. This spectral profile was not new in the EPR literature, where it had been attributed in general to the electron mobility [136–141]. However, electrons in $Ce_{1-x}Gd_xO_{2-x/2}$ have scarce or null mobility at $x \geq 0.2$ [142]. Therefore, combining ESR and atomic force microscopy (AFM) measurements on selected samples, a different explanation was found for this Dysonian profile. Indeed, this spectral shape was attributed to the formation of surface polaritons. The last are propagating coupled electromagnetic–electric dipole excitation modes, created by the diffraction of the microwave ESR field on the surface roughness of the sample, and able to mix absorption and dispersion ESR modes [143]. The formation of polaritons could be further favored by the high spin–orbit interaction characterizing the surface-doping Gd ions. Indeed, AFM data showed the presence of 1.11 nm corrugations on the sample surface when $x = 0.125$, increasing to 1.81 nm with the $x = 0.5$ sample and up to 2.2 with $x = 0.875$. Correspondingly, the Dysonian shape was fitting increasingly better the experimental ESR spectrum of samples with increasing x, up to the perfect fits reported for the spectrum of the $x \geq 0.5$ samples.

4 Spin Exchange Along Nanosized Chains

4.1 Spin Exchange Along –O–Cu–O–Cu– Chains

We have above reported the case in which Cu^{2+} ions were paramagnetic isolated defects doping nanosized diamagnetic particles (paragraph 2.1). Now we will take into consideration micro-sized particles in which Cu–O chains form ordered nanosized sub-systems.

$REBa_2Cu_3O_{6+\delta}$ (RE = Y and Lanthanides; $0 \leq \delta \leq 1$) are antiferromagnetic charge transfer (CT) insulators for $\delta \approx 0$, while they show metallic-like conduction and undergo a superconductive transition [144] at $T_C \approx 90$ K for $\delta \approx 1$. Here, the case of RE = Sm is considered. Oxidation in the O4 site (see Fig. 22) introduces holes in the valence band(s), following equation

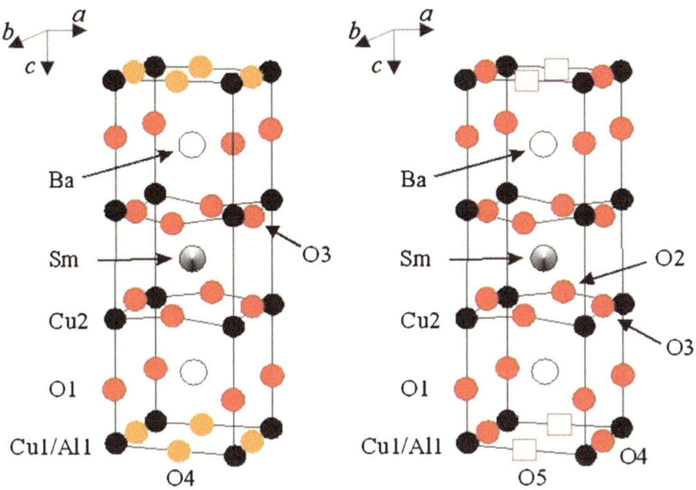

Fig. 22 The crystal structure of $SmBa_2Cu_{3-x}Al_xO_{6+\delta}$. On the *left-hand side* is shown the tetragonal *P4/mmm* structure. O4 sites are only partially occupied. On the *right-hand side* the orthorhombic *Pmmm* structure of $SmBa_2Cu_{3-x}Al_xO_7$ is shown. *Squares* in positions O5 indicate oxygen vacancies

$$\frac{1}{2}O_2 + V_i \rightleftarrows O_i'' + 2h^{\bullet} \tag{19}$$

O_i'' (in the O4 site) is considered an interstitial atoms when the $SmBa_2Cu_3O_6$ compound is taken as a reference.

The increased conductivity upon increasing δ values is due to the charge transfer (CT) of a fraction of the holes concentration from Cu1-O4 chains, where oxidation occurs to the superconducting Cu2-O2/O3 planes [145] (labels following Ref. [146] as in Fig. 22).

Oxidation causes a progressive ordering of Cu1-O4 ions along the crystallographic *b* direction, which in turn causes a structural phase transition from *P4/mmm* space group to orthorhombic *Pmmm* space group. Moreover, internal pressure is introduced by inter-atomic distance changes [147]. Thus, in order to understand the charge transfer mechanism, a separation of the above effects was needed.

In order to highlight the specific role of Cu–O chains length and ordering on the rising of superconductivity in $SmBa_2Cu_{3-x}Al_xO_{6+\delta}$ compound, samples with the same oxygen and with the same Al doping amounts, but different mean Cu–O chains' lengths, have been investigated [149]. In fact, aluminum substitutes copper only at the Cu1 sites. Therefore, taking fixed the oxygen concentration δ, Al doping at the Cu1 site implies a decrease of the hole concentration in the compound because of the different charges of the two ions.

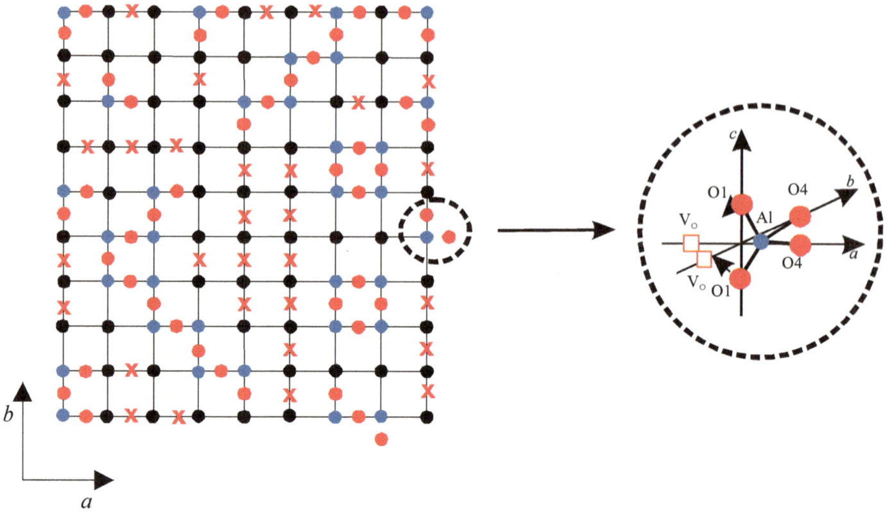

Fig. 23 *Black*, *blue* and *red circles* represent Cu1, Al and O4 ions in reduced conditions; *red crosses* stay for the O ions introduced by oxidation. The *inset* displays the tetrahedric environment for Al

A further effect of Al substitution for Cu1 on the structure [148] is to cut the Cu1-O4 chains. This is due to the Al tetrahedric coordination [148–150], contrasting the Cu1 square planar coordination (Fig. 23).

Two $SmBa_2Cu_{3-x}Al_xO_{6+\delta}$ samples, A and B, with the same x and (almost) δ values, have been compared by EPR spectroscopy. In particular, δ was slightly larger for A than for B implying a larger hole concentration for the former sample. Due to two different annealing routes, Al ions were more stochastically distributed in the A sample and more clusterized in the B one. Clustering assists the formation of Cu1-O4 chains longer in the latter than in the former.

This different chain mean length affects also the EPR spectra, as shown in Fig. 24. At $T = 280$ K these patterns were rather similar for the two compared samples, being composed of a Lorentzian-shaped line (L) at $g \cong 2$, to which a broad, low-intensity Gaussian-shaped line (G) added at lower magnetic field. At lower temperature, however, G became more intense and broader, also moving toward lower field values, these effects being more accentuated for A than for B. Furthermore, L was markedly increasing at lower temperature only with the latter.

A Lorentzian shape was indicating the occurrence of a spin–spin exchange in both samples. Indeed, units like

$$-Cu^{2+}\downarrow -O^{2-}\uparrow\downarrow -Cu^{2+}\uparrow -O^{-}\uparrow -Cu^{2+}\uparrow -O^{2-}\uparrow\downarrow - \text{(Scheme I)}$$

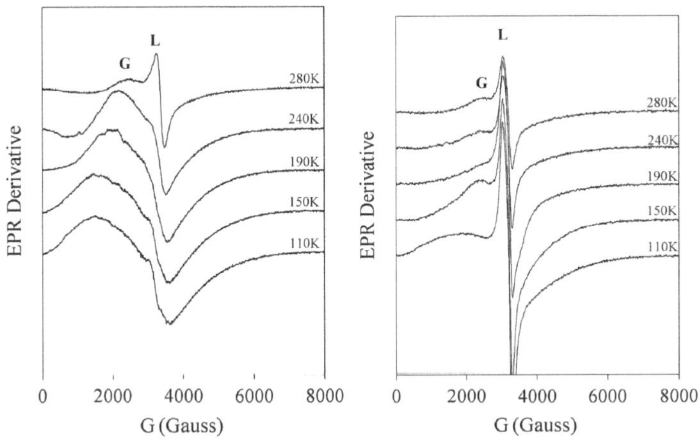

Fig. 24 EPR spectra corresponding to the A sample (*left*) and to the B sample (*right*) at different temperatures. G and L refer to the Gaussian- and Lorentzian-type peaks, respectively (modulation amplitude 4 gauss, microwave power 20.07 mW, gain 60 dB). Reprinted with permission from Ref. [149]

are present in the Al-undoped oxidized $SmBa_2Cu_3O_7$ sample, both in the Cu1-O4 and in the Cu2-O2/3 planes. The three unpaired spins on the right in Scheme I are coupled ferromagnetically with one other [151] owing to the orthogonality of $p\pi$ and d orbitals, occupied by the unpaired electrons in O and in Cu holes, respectively. This ferromagnetic coupling overcomes always [152, 153] the superexchange interaction occurring between the central and the left Cu ions in Scheme I. Because of the hopping of an O-$p\pi$ hole [151, 152], ferromagnetically coupled Cu^{2+} ions swap continuously with those coupled antiferromagnetically, giving rise to the narrowed L feature.

In the Cu1-O4 plane, Al doping would produce fragments like

$$-Al^{3+}_{Cu1} - O^{2-}_{O4} \uparrow\downarrow - Cu^{2+}_{Cu1} \downarrow - O^{2-}_{O4} \uparrow\downarrow - Cu^{2+}_{Cu1} \uparrow - O^{2-}_{O4} \uparrow\downarrow - \text{(Scheme II)}$$

which no more contribute to EPR. The dispersion of Al in the Cu1-O4 plane (as with A) generates spin bags (SBs), i.e., disordered systems made up of different spin-polarized clusters of O^- holes with limited mobility [153, 154], generating the ferromagnetic (FM) G pattern [110] shown in Fig. 24(left). The SB size decreases with increasing temperature, causing a spectral drift toward the $g \cong 2$ region accompanied by lower intensity. However, Al clustering in B induces a pronounced increase of the L/G ratio, indicating that a more ordered system forms, in which SBs almost disappear while longer chains like those of Scheme I form. Therefore, mean mobility of conduction holes and CT increases in B, in spite of the slightly higher total amount of holes in A.

The DC zero-field-cooled and field-cooled susceptibility measurements showed that Sample A was characterized by $T_C = 46.8$ K, lower than for Sample B ($T_C = 56.7$ K) and by a limit of the Meissner fraction of ~9 %, lower than for the latter (~13 %).

In spite of the slightly higher total amount of holes in A with respect to B ($\delta_A > \delta_B$), the CT to the Cu2-O2/3 superconducting planes improves in B by increasing the mean chain length as well as T_C and Meissner fraction. Therefore, it was concluded that local Cu1-O4 ordering is fundamental in determining the amount of CT and the superconducting properties in $REBa_2Cu_3O_{6+\delta}$ system.

4.2 *Ferromagnetic and Antiferromagnetic Interactions Along –O–Co–O–Co– Chains*

Perovskitic samples of formula $La_{1-x}Ce_xCoO_3$ are well known as quite active catalysts for exhaust gas depollution [155, 156] as well as for the catalytic flameless combustion (CFC) of methane [157, 158]. Indeed, samples of this kind, prepared by the traditional sol-gel (SG) procedure, have been successfully tested for CO oxidation with air [155] and NO reduction by CO [156].

The $3d^6$ Co^{3+} ions, in general, are stable in the low spin ($S = 0$, t_{2g}^6, e_g^0) state when octahedrally coordinated to six oxygen atoms in an ideal $LaCoO_3$ structure [159], so that no ESR spectrum was expected with these samples. However, a $g \cong 2$ Lorentzian-shaped line (see Fig. 25) was revealed [158] by $La_{1-x}Ce_xCoO_3$ with a $x = 0.05$ sample prepared by calcinations of sol-gel citrate precursors (hereafter CN) as well as with samples of composition $x = 0.05$ and 0.2 prepared from oxide mixtures in molten ammonium nitrate (hereafter AN). Those ESR features were broadening with increasing temperature.

The substitution of Ce^{4+} for La^{3+} would reduce some cobalt ions to Co^{2+}. The last can be either in the ($S = ½$) low spin state or in a ($S = 3/2$) high-spin state. However, the latter is characterized by very fast relaxation times, giving [159] ESR spectrum at temperature lower than ~20 K only. Therefore, only the ($S = ½$) low spin Co^{2+} ions are generally detectable. These ions, when octahedrally coordinated to oxygen, could form $–Co^{2+}–O^{2-}–Co^{2+}–O^{2}–$ chains in which the paramagnetic Co^{2+} ions would be coupled antiferromagnetically to each other through superexchange mediated by the O^{2-} ions. The presence of oxygen vacancies interrupting these chains could account for the observation of the Lorentzian-shaped line [156, 160]. However, the last was no more observed with the $x = 0.2$ CN sample which, by the way, was the best catalyst for the CFC of methane. This ESR feature could be attributed to the presence of Co_3O_4 phase impurities, though the XRD patterns seemed to show this phase in the $x = 0.2$ samples of both A and C series [158]. Therefore, the attribution of this line can be considered rather controversial, on the base of these data only.

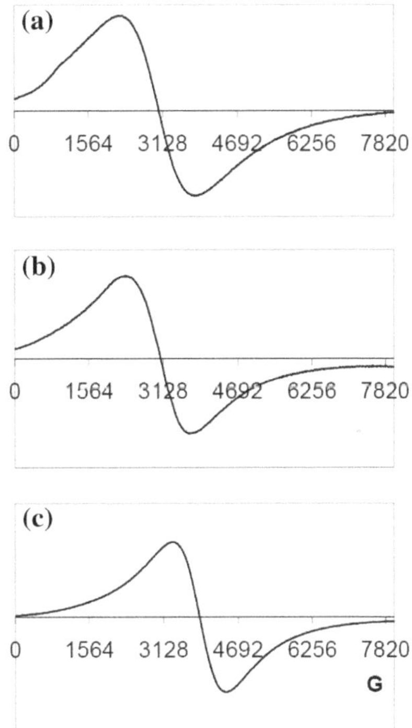

Fig. 25 Some Lorentzian-shaped EPR lines detected at 140 K. **a** $x = 0.05$ CN; **b** $x = 0.05$ AN; and **c** $x = 0.2$ AN. ΔH_{pp} = **a** 1580 G; **b** 1360 G, **c** 1060 G. Adapted with permission from Ref. [158]

Therefore, $La_{1-x}Ce_xCoO_3$ samples prepared by a different method (flame hydrolysis, FH) were analyzed. They were characterized by a perfect perovskite structure, though the last is not thermodynamically favored with respect to other Co oxidic species. In addition, these FH samples did not show the above-mentioned ESR spectrum [157]. Then, in a bit more recent investigation [161] a few further $La_{1-x}Ce_xCoO_3$ ($x = 0.1$) catalysts were prepared by the traditional sol-gel (SG) or by flame pyrolysis (FP) [162] procedures and analyzed by ESR spectroscopy at $120 \leq T \leq 300$ K. Small amounts of Co_3O_4 and CeO_2 phases were noticed in the $x = 0.1$ (SG1) catalyst, accompanied by La_2O_3. This sample gave a single very faint ESR feature at $g \cong 2$, Lorentzian shaped at 120 K and broadening with increasing temperature, as in the previous cases. By contrast, the single-phase $x = 0.1$ (FP1) sample did not give any ESR spectrum.

We can conclude that the pure mono-phasic perovskite-like $La_{1-x}Ce_xCoO_3$ samples did not show any $g \cong 2$ Lorentzian-shaped line, as indeed observed with FP1 and with the samples prepared by the FH method [157], whereas this ESR feature was attributed to (presumably nanosized) particles of Co_3O_4, when forming in these systems. Indeed, a $g \cong 2$ Lorentzian-shaped line, broadening with increasing temperature, has been noticed since a long time with Co_3O_4 [160, 163, 164]. Narrower lines were obtained at higher preparation temperature, i.e., at lower residual disorder. The magnetic structure of this cobalt oxide has been deeply investigated [164]. Units like

- O^{2-} - Co^{3+} - O^{2-} - Co^{2+} - O^{2-} - Co^{3+} - O^{2-} - Co^{2+} - O^{2-} - (Scheme III)

have been hypothesized, in which two Co^{2+} ($3d^7$, S = 3/2 or 1/2) ions, localized in tetrahedral sites, would interact with each other through O^{2-}–Co^{3+}–O^{2-} bridges, Co^{3+} ($3d^6$, S = 0) being localized in octahedral field. Two sets of units like that outlined in Scheme III would form, characterized by antiferromagnetic (AFM) and ferromagnetic (FM) coupling, respectively. The former would have multiplicity 12, and the latter multiplicity 24. Other similar schemes have been elsewhere proposed [160, 163]. In any case, a prevailing of FM over AFM interactions in SG1 was suggested.

The line broadening with increasing temperature was interpreted on the base of *polaron* propagation, i.e., of spin–phonon interactions.

It should be noted that in this case, a "negative" correlation between ESR signal and catalytic activity was highlighted. ESR silent (active) were good (bad) catalysts. In fact, the segregation of CeO_2 (as well as of Co_3O_4 which generate the ESR signal) hindered the formation of Co^{2+}/Co^{3+} couples involved in the catalytic process.

4.3 Polaron Mobility Along –O–Mn–O–Mn–O– Nanosized Chains

$La_{1-x}M_xMnO_{3+\delta}$ (M = Ce, Sr) mixed oxides have been deeply investigated in the recent past. Among the possible practical applications, there is the catalytic flameless combustion (CFC) of hydrocarbons [165]. To this purpose, it is compulsory to combine their thermal stability to their high specific surface area (SSA). These and other properties of doped manganites strongly depend on the procedure followed in their preparation [166]. In particular, the synthesis parameters affect the incorporation degree of the ion M, substituting for La [167, 168].

The last parameter was particularly important because the partial substitution of Sr^{2+} for La^{3+} was favoring the oxidation of some Mn^{3+} to Mn^{4+}, increasing the catalytic activity of the sample for the CFC of methane [169, 170] and giving rise to a mixed valence compound which was displaying a variety of phenomena such as electron conduction linked to double exchange interactions between Mn^{3+} and Mn^{4+} species along chains like

…–O-Mn^{3+}-O-Mn^{4+}-O-… (scheme IV)

[171, 172] and ferromagnetic transition due to ferromagnetic polarization arising around the Mn^{4+} ions at low T [172–174].

A different case was that of Ce^{4+} substituting for La^{3+}, in which some Mn^{3+} might reduce to Mn^{2+} ions [172], which contribute to the ESR pattern (Fig. 26a). However, in the last cited paper, a single EPR line was observed also with the undoped as-prepared sample. This fact was considered rather intriguing, as Mn^{3+} (if not involved in a double exchange with Mn^{4+}) is characterized by a very fast relaxation time, so that its ESR spectrum is detectable at very low temperature only. Indeed, in the undoped sample this line could be attributed to Mn^{3+} ions only, in the presence of a strong coupling between lattice, electronic, and magnetic degrees of freedom due to static Jahn–Teller (JT) effect. The last would distort the oxygen octahedra of the cubic cell giving rise to an orthorhombic or monoclinic [175] structure, so that the degeneracy would be removed by lowering the energy of the occupied d_{3x2-r2} or d_{3y2-r2} orbital. Then, the Mn^{3+} ions would acquire a zig-zag-type ordering of occupied orbitals in *ab* planes. In this situation, an ESR line was expected at about $g_z \cong 1.95$ and $g_x = g_y = 1.99$, in agreement with experimental literature data [159, 173]. No variation of the ESR peak-to-peak linewidth was expected for this line at the detected temperatures, ranging between 280 and 550 K, because $LaMnO_3$ remains JT distorted up to T_{JT} = 750 K [176]. By contrast, unexpectedly, the EPR line was broadening with increasing temperature in [172]. Therefore, it was not possible to explain this effect by just invoking the static JT effect. Some temperature-dependent spin–spin relaxation process might be also hypothesized. This was singled out as the temperature-dependent interactions between phonons and conduction electrons (i.e., to polarons formation) [172, 177].

Referring again to the catalytic properties, the materials described above, prepared using the traditional sol-gel (SG) preparation, did not get sufficient thermal resistance. Therefore, an alternative preparation procedure by flame pyrolysis (FP) [162, 178, 179] was proposed for $La_{1-x}M_xMnO_{3+\delta}$ ($x = 0$, $x \cong 0.1$ with M = Ce; $x \cong 0.1$ and $\cong 0.2$ with M = Sr). The FP preparation of undoped and Ce-doped samples was leading to nanoparticles with monoclinic structure, whereas Sr-doped catalysts showed orthorhombic/rhombohedral symmetry. The observed structural phases were connected to the concentration of Mn^{4+} present in the Sr-doped samples, and to the fact that the undoped and the Ce-doped monoclinic samples were almost Mn^{4+} free [136].

Samples prepared by this way have been compared to sample prepared by a traditional SG process by examining their ESR spectra and X-ray powder diffraction patterns [136]. ESR measurements were carried out on samples (i) as prepared, (ii) after activation, and (iii) after reaction. The single ESR line was observed also with the undoped FP-prepared $LaMnO_{3+\delta}$ sample [136] (Fig. 26a). Its peak-to-peak linewidth variation in the $250 \leq T \leq 420$ K range is shown in Fig. 26b. A marked ESR line narrowing was always observed after the activation treatment, similar to that noticed with the same samples after their catalytic reaction (see for example inset of Fig. 26a). We note that the spectrum of the as-prepared one (dotted line in Fig. 26a) displays an exchange-narrowed ESR feature with a bit asymmetric shape, well described by the Dysonian model. As above reported in the paragraph 2.4, the Dysonian lineshape can be attributed to an α dispersion-to-absorption contribution to the Lorentzian-shaped line. This effect

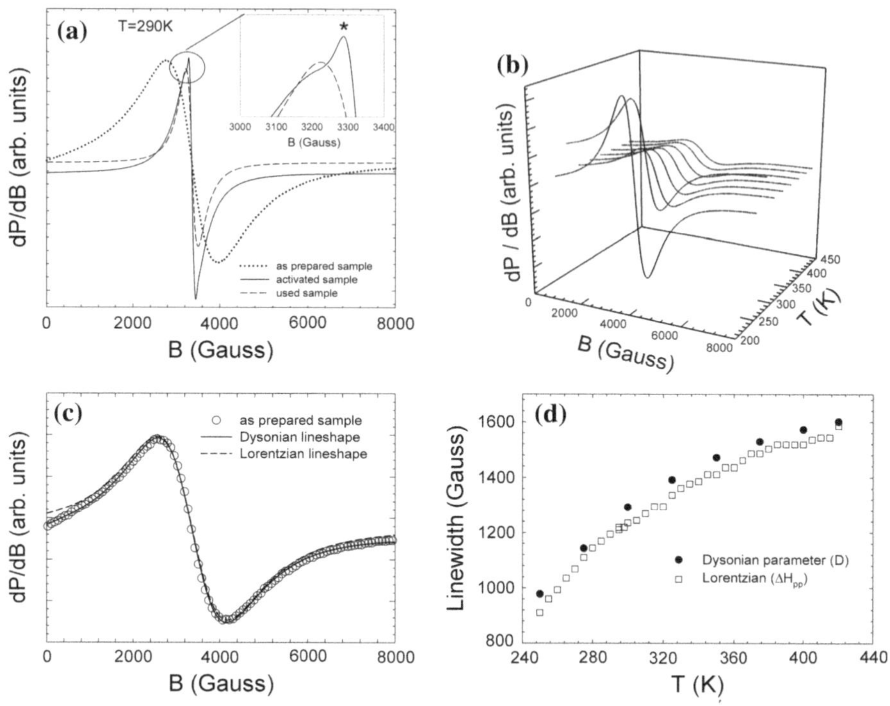

Fig. 26 ESR spectra of $LaMnO_3$. **a** ESR spectra of as-prepared sample (*dotted line*), after activation (*solid line*), after reaction (*dashed line*) at T = 290 K. (*) Mn^{2+} ESR contribution. **b** ESR spectra of the as-prepared sample as a function of temperature; **c** ESR spectra of as-prepared sample collected at T = 290 K (*empty dots*) fitted using a Dysonian (*solid line*) and Lorentzian (*dotted line*) functions; **d** Comparison between the parameters obtained by the Lorentzian model (ΔH_{pp}) and the Dysonian parameter (*D*) (see text for details). Reprinted with permission from Ref. [136]. Copyright (2012) American Chemical Society

occurs in samples in which the electron mobility is high enough to create the so-called "skin effect" for the microwave field employed by the ESR spectrometer [180, 181]. A $\alpha = 1$ value would correspond to dispersion of the same strength of absorption, in a sample larger than its skin depth. On the opposite, in any case $\alpha = 0$ with samples smaller than their skin depth. The situation reported in [136] was an intermediate one, with $0 < \alpha < 1$.

The Dysonian and the Lorentzian lineshape models for the ESR spectrum of undoped FP-prepared $LaMnO_3$ were compared (Fig. 26c). Only a little discrepancy was noticed in the low-field part of the line, so that the T_2 spin–spin relaxation time values obtained through these two models were very close to each other (Fig. 26d). Therefore, the peak-to-peak linewidth obtained by the Lorentzian model, being $\Delta H_{pp} = (2/\sqrt{3})/T_2$, was retained valid in order to compare quantitatively the T_2 values at different doping amounts and with samples prepared by the different SG and FP procedures (see Fig. 27).

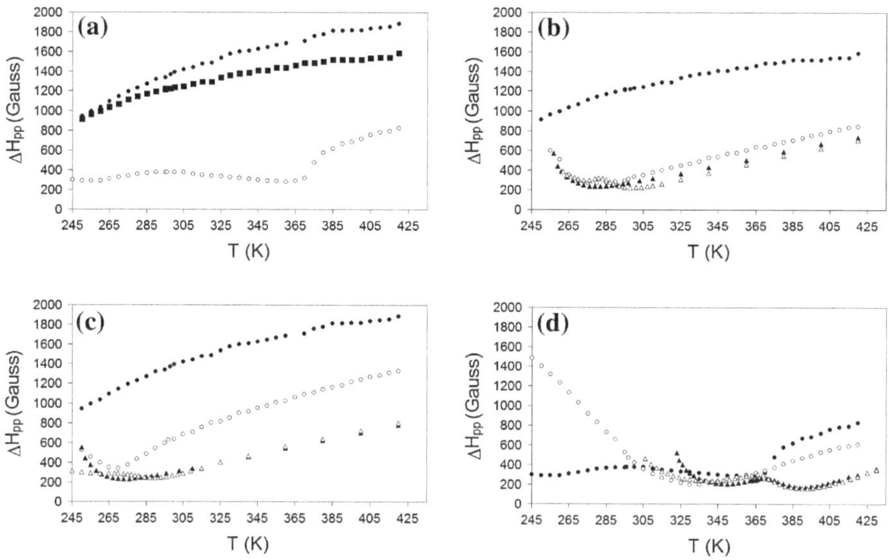

Fig. 27 Temperature dependences of peak-to-peak width (ΔH_{pp}) of the $g \cong 2$ ESR line observed. The ΔH_{pp} values refer to **a** $La_{0.9}Ce_{0.1}MnO_3$ (*full circles*), $LaMnO_3$ (*full squares*), $La_{0.9}Sr_{0.1}MnO_3$ (*empty circles*) FP as-prepared samples; **b** $LaMnO_3$–FP as-prepared (*full circles*) and used (*empty circles*) samples and $LaMnO_3$–SG as-prepared (*full triangles*) and used (*empty triangles*) samples; **c** $La_{0.9}Ce_{0.1}MnO_3$–FP as-prepared (*full circles*) and used (*empty circles*) samples and $La_{0.9}Ce_{0.1}MnO_3$–SG as-prepared (*full triangles*) and used (*full triangles*) samples; **d** $La_{0.9}Sr_{0.1}MnO_3$–FP as-prepared (*full circles*) and used (*empty circles*) samples and $La_{0.9}Sr_{0.1}MnO_3$–SG as-prepared (*full triangles*) and used (*empty triangles*) samples. Reprinted with permission from Ref. [136]. Copyright (2012) American Chemical Society

The ESR line broadens with increasing temperature a bit more markedly with the Ce-doped than with the undoped sample (Fig. 27a). This has been attributed to the formation of some electron–hole (EH) pairs [182–184] composed of Mn^{2+}–Mn^{4+} ions. Indeed, EH pairs have a very large fluctuating electrical dipole moment, able to cause attractive electrostatic dipole–dipole interactions. Therefore, EH pairs can group themselves into an EHBL (Bose Liquid) phase. The formation of the last would subtract some Mn^{4+} ions from the DE, decreasing its ESR line-narrowing efficiency. On the other hand, Mn^{3+} ions are by far more abundant than the Mn^{4+} ones, so that a decreased amount of the latter only would be sufficient to cause a decreased DE.

By contrast, doping $LaMnO_3$ by Sr^{2+} was increasing the amount of the Mn^{4+} ions present in the material. This was confirmed by the ESR lineshape, which was by far narrower with the $x = 0.1$ Sr-doped sample than with the undoped and Ce-doped ones (Fig. 27a). This strong narrowing could be attributed to an increased Mn^{4+}–O–Mn^{3+} DE, favored by the greater availability of Mn^{4+} ions [172, 174], as well as by the smaller distortion of the Mn–O–Mn bond angle arising from the substitution of Sr^{2+} for La^{3+} [185].

The undoped $LaMnO_3$ sample showed an ESR line broader with FP-prepared than with SG-prepared samples (see Fig. 27b, where the ΔH_{pp} values obtained after reaction are also reported, though discussed on the original paper, only) [172, 177]. We can generalize that the ESR line of the nanosized FP-prepared samples was about 2–3 times (1.3 times with M = Sr) broader than those of the analogous SG-prepared materials, at least in the range of $\sim$280 (380 with M = Sr) $< T <$ 420 K.

However, we must outline that ESR features broader than with SG samples were already reported and discussed for flame hydrolysis (FH) prepared undoped La manganite at room temperature [177]. The observed differences in ESR linewidth can be related to different structural sample phases. Indeed, the undoped manganite was rhombohedral when SG-prepared, orthorhombic when FH-prepared, and monoclinic in the present FP-prepared sample [177].

In addition, the linewidth of undoped and of Ce-doped samples was continuously increasing with increasing temperature when the samples were prepared by the FP procedure (Fig. 27a), but not when they were prepared by the classical SG method (full triangles in Fig. 27b, c). In the latter case, the linewidth versus temperature revealed a relative minimum at $T = T_{min}$, and a line broadening at $T < T_{min}$ accompanied by increased spectral intensity, typical in the presence of a magnetic transition to a ferromagnetic state [172]. Indeed, the undoped and Ce-doped $LaMnO_3$ samples were characterized by an abundance of Mn^{4+} ions lower when in the monoclinic phase obtained by the FP procedure than when in the rhombohedral phase obtained by the SG procedure. Therefore, a lower amount of free electrons available to form FM domains was present in the former than in the latter case. The complex relationship between crystallographic structure and ESR as a function of dopant concentration and annealing conditions is described in term of Mn^{4+}/Mn^{3+} ratio in [136].

References

1. Chen, X., Mao, S.S.: Titanium dioxide nanomaterials: synthesis, properties, modifications, and applications. Chem. Rev. **107**, 2891–2959 (2007)
2. Fujishima, A., Zhang, X., Tryk, D.: TiO_2 photocatalysis and related surface phenomena. Surf. Sci. Rep. **63**, 515–582 (2008)
3. Meroni, D., Ardizzone, S., Cappelletti, G., Ceotto, M., Ratti, M., Annunziata, R., Benaglia, M., Raimondi, L.: Interplay between chemistry and texture in hydrophobic TiO_2 hybrids. J. Phys. Chem. C **115**, 18649–18658 (2011)
4. Taylor, S., Mehta, M., Samakhvalov, A.: Production of hydrogen by glycerol photoreforming using binary nitrogen-metal-promoted N-M-TiO_2 photocatalysts. ChemPhysChem **15**, 942–949 (2014)
5. Castro, A.L., Nunes, M.R., Carvalho, M.D., Ferreira, L.P., Jumas, J.C., Costa, F.M., Florencio, M.H.: Doped titanium dioxide nanocrystalline powders with high photocatalytic activity. J. Solid State Chem. **182**, 1838–1845 (2009). (b) Cottineau, T., Béalu, N., Gross, P.-A., Pronkin, S.N., Keller, N., Savinova, E.R., Keller, V.: One step synthesis of niobium

doped titania nanotube arrays to form (N,Nb) co-doped TiO2 with high visible light photoelectrochemical activity. J. Mater. Chem. A **1**, 2151 (2013)
6. Kudo, A., Niishiro, R., Iwase, A., Kato, H.: Effects of doping of metal cations on morphology, activity, and visible light response of photocatalysts. Chem. Phys. **339**, 104–110 (2007)
7. Furubayashi, Y., Hitosagi, T., Yamamoto, Y., Inaba, K, Kinoda, G., Hirose, Y., Shimada, T., Hasegawa, T.: Transparent metal: Nb doped Anatase TiO2. Appl. Phys. Lett. **86**, 252101 (2005). (b) Hitosugi, T., Yamada, N., Nakao, S., Hirose, Y., Hasegawa, T.: Properties of TiO2-based transparent conducting oxides. Phys. Status Solidi A **207**, 1529–1537 (2010)
8. Lü, X., Mou, X., Wu, J., Zhang, D., Zhang, L., Huang, F., Xu, F., Huang, S.: Improved-performance dye-sensitized sola cell using Nb-doped TiO_2 electrodes: efficient electron injection and transfer. Adv. Funct. Mater. **20**, 509–515 (2010)
9. Chandiran, A.K., Sauvage, F., Casas-Cabanas, M., Comte, P., Zakeeruddin, S.M., Grätzel, M.: Doping a TiO_2 photoanode with Nb^{5+} to enhance transparency and charge collection efficiency in dye-sensitized solar cells. J. Phys. Chem. C **114**, 15849–15856 (2010)
10. Choudhury, B., Munmun Dey, M., Choudhury, A.: Defect generation, *d-d* transition, and band gap reduction in Cu-doped TiO_2 nanoparticles. Int. Nano Lett. **3**, 25 (2013)
11. Atherton, N.M.: Electron Spin Resonance, Chapter 6. Ellis Horwood Ltd, John Wiley (1973)
12. Zhang, J., Wu, Y., Xing, M., Leghari, S.A.K., Sajjad, S.: Development of modified N doped TiO_2 photocatalyst with metals, nonmetals and metal oxides. Ener. Environ. Sci. **3**, 715–726 (2010)
13. Ihara, T., Miyoshi, M., Iriyama, Y., Matsumoto, O., Sigihara, S.: Visible-light-active titanium oxide photocatalyst realized by an oxygen-deficient structure and by nitrogen doping. Appl. Catal. B: Environ. **42**, 403–409 (2003)
14. Sato, S., Nakamura, R., Abe, S.: Visible-light sensitization of TiO_2 photocatalysts by wet-method N doping. Appl. Catal. A **284**, 131–137 (2005)
15. Spadavecchia, F., Cappelletti, G., Ardizzone, S., Ceotto, M., Azzola, M.S., Lo Presti, L., Cerrato, G., Falciola, L.: Role of Pr on the Semiconductor Properties of Nanotitania. An Experimental and First-Principles Investigation. J. Phys. Chem. C **116**, 23083–23093 (2012)
16. Livraghi, S., Viotta, A., Paganini, M.C., Giamello, E.: The nature of paramagnetic species in nitrogen doped TiO_2 active in visible light photocatalysis. Chem. Commun. 498–500 (2005)
17. Di Valentin, G., Pacchioni, G., Selloni, A., Livraghi, S., Giamello, E.: Characterization of paramagnetic species in N-Doped TiO_2 powders by EPR spectroscopy and DFT calculations. J. Phys. Chem. B **109**, 11414–11419 (2005)
18. Joung, S.-K., Amamiya, T., Murabayashi, M., Itoh, K.: Relation between photocatalytic activity and preparation conditions for nitrogen-doped visible light-driven TiO_2 photocatalysts. Appl. Catal. A: General **312**, 20–26 (2006)
19. Livraghi, S., Paganini, M.C., Giamello, E., Selloni, A., Di Valentin, C., Pacchioni, G.: Origin of photoactivity of nitrogen-doped titanium dioxide under visible light. J. Am. Chem. Soc. **128**, 15666–15671 (2006)
20. Serpone, N.: Is the band gap of pristine TiO_2 Narrowed by anion- and cation-doping of titanium dioxide in second-generation photocatalysts? J. Phys. Chem. B **110**, 24287–24293 (2006)
21. Di Valentin, C., Finazzi, E., Pacchioni, G., Selloni, A., Livraghi, S., Paganini, M.C., Giamello, E.: N-doped TiO_2: theory and experiment. Chem. Phys. **339**, 44–56 (2007)
22. Reyes-Garcia, E.A., Sun, Y., Reyes-Gil, K., Raftery, D.: ^{15}N solid state NMR and EPR characterization of N-Doped TiO_2 photocatalysts. J. Phys. Chem. C **2007**(111), 2738–2748 (2007)
23. Sun, H., Bai, Y., Lin, H., Lin, W., Xu, N.: Photocatalytic decomposition of 4-chlorophenol over an efficient N-doped TiO_2 under sunlight irradiation. J. Photochem. Photobiol. A: Chem. **201**, 15–22 (2009)
24. Emeline, A.V., Kuznetsov, V.N., Rybchuk, V.R., Serpone, N.: Visible-light-active titania photocatalysts: The case of N-doped TiO_2s—properties and some fundamental issues. Intern. J. Photoenergy **2008**, ID 258394, 1–19 (2008)

25. Bianchi, C.L., Cappelletti, G., Ardizzone, S., Gialamella, S., Naldoni, A., Oliva, C., Pirola, C.: N-doped TiO_2 from $TiCl_3$ for photodegradation of air pollutants. Catal. Today **144**, 31–36 (2009)
26. Spadavecchia, F., Cappelletti, G., Ardizzone, S., Bianchi, C.L., Cappelli, S., Oliva, C., Scardi, P., Leoni, M., Fermo, P.: Solar photoactivity of nano-N-TiO_2 from tertiary amine: role of defects and paramagnetic species. Appl. Catal. B: Environ. **96**, 314–322 (2010)
27. Meroni, D., Ardizzone, S., Cappelletti, G., Ceotto, M., Ratti, M., Annunziata, R., Benaglia, M., Raimondi, L.: Interplay between chemistry and texture in hydrophobic TiO_2 hybrids. J. Phys. Chem. C **115**, 18649–18658 (2011)
28. Barolo, G., Livraghi, S., Chiesa, M., Paganini, M.C., Giamello, E.: Mechanism of the photoactivity under visible light of N-doped titanium dioxide. Charge carriers migration in irradiated N-TiO2 investigated by electron paramagnetic resonance. J. Phys. Chem. C **116**, 20887–20894 (2012)
29. Spadavecchia, F., Ardizzone, S., Cappelletti, G., Oliva, C., Cappelli, S.: Time effects on the stability of the induced defects in TiO_2 nanoparticles doped by different nitrose sources. J. Nanoparticles Res. **14**, 1301 (1–12) (2012)
30. Marchiori, C., Di Liberto, G., Soliveri, G., Loconte, L., Lo Presti, L., Meroni, D., Ceotto, M., Oliva, C., Cappelli, S., Cappelletti, G., Aieta, C., Ardizzone, S.: Unraveling the cooperative mechanism of visible-light absorption in bulk N,Nb codoped TiO_2 powders of nanomaterials. Phys. Chem. C **118**, 24152–24164 (2014)
31. Lo Presti, L., Ceotto, M., Spadavecchia, F., Cappelletti, G., Meroni, D., Acres, R.G., Ardizzone, S.: Role of the nitrogen source in determining structure and morphology of N-doped nanocrystalline TiO_2. J. Phys. Chem. C **118**, 4797–4807 (2014)
32. Yates, H.M., Nolan, M.G., Sheel, D.W., Pemble, M.E.: The role of nitrogen doping on the development of visible light-induced photocatalytic activity in thin TiO_2 films grown on glass by chemical vapour deposition. J. Photochem. Photobiol., A **179**, 213–223 (2006)
33. Ma, X., Wu, Y., Lu, Y., Xu, J., Wang, Y., Zhu, Y.: Effect of compensated codoping on the photoelectrochemical properties of anatase TiO_2 photocatalyst. J. Phys. Chem. C **115**, 16963–16968 (2011)
34. Zhao, Z., Liu, Q.: Designed highly effective photocatalyst of anatase TiO_2 codoped with nitrogen and vanadium under visible-light irradiation using first-principles. Catal. Lett. **124**, 111–117 (2008)
35. Gai, Y., Li, J., Li, S.-S., Xia, J.-B., Wei, S.-H.: Design of narrow-gap TiO_2: a passivated codoping approach for enhanced photoelectrochemical activity. Phys. Rev. Lett. **102**, 036402 (2009)
36. Kuvarega, A.T., Krause, R.W.M., Mamba, B.B.: Nitrogen/palladium-codoped TiO_2 for efficient visible light photocatalytic dye degradation. J. Phys. Chem. C **115**, 22110–22120 (2011)
37. Wang, X., Tang, Y., Leiw, M.-Y., Lim, T.-T.: Solvothermal synthesis of Fe–C codoped TiO_2 nanoparticles for visible-light photocatalytic removal of emerging organic contaminants in water. Appl. Catal. A **409–410**, 257–266 (2011)
38. Thind, S.S., Wu, G., Chen, A.: Synthesis of mesoporous nitrogen–tungsten co-doped TiO_2 photocatalysts with high visible light activity. Appl. Catal. B **111–112**, 38–45 (2012)
39. Marquez, A.M., Plata, J.J., Ortega, Y., Sanz, J., Colon, G., Kubacka, A., Fernandez-Garcia, M.: Making photo-selective TiO_2 materials by cation-anion codoping: from structure and electronic properties to photoactivity. J. Phys. Chem. C **116**, 18759–18767 (2012)
40. Gu, D.-E., Yang, B.-C., Hu, Y.-D.: V and N co-doped nanocrystal anatase TiO_2 photocatalysts with enhanced photocatalytic activity under visible light irradiation. Catal. Commun. **9**, 1472–1476 (2008)
41. Cong, Y., Zhang, J., Chen, F., Anpo, M., He, D.: Preparation, photocatalytic activity, and mechanism of nano-TiO_2 co-doped with nitrogen and Iron (III). J. Phys. Chem. C **111**, 10618–10623 (2007)

42. Breault, T.M., Bartlett, B.M.: Lowering the band gap of anatase-structured TiO_2 by coalloying with Nb and N: electronic structure and photocatalytic degradation of methylene blue dye. J. Phys. Chem. C **116**, 5986–5994 (2012)
43. Breault, T.M., Bartlett, B.M.: Composition dependence of TiO2:(Nb, N)-x compounds on the rate of photocatalytic methylene blue dye degradation. J. Phys. Chem. C **117**, 8611–8618 (2013)
44. Cottineau, T., Béalu, N., Gross, P.-A., Pronkin, S.N., Keller, N., Savinova, E.R., Keller, V.: One step synthesis of niobium doped titania nanotube arrays to form (N, Nb) co-doped TiO2 with high visible light photoelectrochemical activity. J. Mater. Chem. A **1**, 2151 (2013)
45. Spadavecchia, F., Cappelletti, G., Ardizzone, S., Ceotto, M., Falciola, L.: Electronic structure of pure and N-doped TiO_2 nanocrystals by electrochemical experiments and first principles calculations. J. Phys. Chem. C **115**, 6381–6391 (2011)
46. Spadavecchia, F., Cappelletti, G., Ardizzone, S., Ceotto, M., Falciola, L.: Electronic structure of pure and N-doped TiO_2 nanocrystals by electrochemical experiments and first principles calculations. J. Phys. Chem. C **115**, 6381–6391 (2011)
47. De Tritio, L., Buonsanti, L., Schimf, A.M., Lordes, A., Gamelin, D.R., Simonutti, R., Milliron, D.J.: Nb-doped colloidal TiO_2 nanocrystals with tunable infrared absorption. Chem. Mater. **25**, 3383–3390 (2013)
48. Mukovozov, I.E., Ezerets, A.M., Vishniakov, A.V., Forni, L., Oliva, C.: Structural effects of interaction between lanthanum cuprate and cerium dioxide. J. Mater. Sci. **32**, 4991–4997 (1997)
49. Oliva, C., Forni, L., Ezerets, A.M., Mukovozov, I.E., Vishniakov, A.V.: EPR characterization of $(CeO_2)_{1-y}$ $(La_2\,CuO_4)_y$ oxide mixtures and their activity for NO reduction by CO. J.C.S. Faraday Trans. **94**, 587–592 (1998)
50. Oliva, C., Cappelli, S., Kryukov, A., Chiarello, G.L., Vishniakov, A.V., Forni, L.: EMR charaterisation of $La_{1.8}M_{0.2}CuO_4$ and $La_{0.9}M_{0.1}CoO_3$ (M = Pr, Sm, Tb) catalysts for methane flameless combustion. J. Mol. Catal. A Chem. **247**, 248–252 (2006)
51. Abragam, A., Bleaney, B.: Electron paramagnetic resonance of transition ions. Oxford University Press, New York (1970)
52. Baker, J.M., Eaton, G.R., Eaton, S.S., Salikhov, K.M. (eds.): Foundations of Modern EPR, Chapter B3. World Scientific, Singapore (1998)
53. Bagguley, D.M., Bleaney, B.: Ferromagnetic resonance at the Clarendon Laboratory, Oxford. A tribute to J. H. E. Griffiths (1908–1981). Contemp. Phys. **31**, 35–42 (1990)
54. Forni, L., Rossetti, I.: Catalytic combustion of hydrocarbons over perovskites. Appl. Catal. B: Environ. **38**, 29–37 (2002)
55. Oliva, C., Cappelli, S., Rossetti, I., Kryukov, A., Bonoldi, L., Forni, L.: Effect of M ion oxidation state in $Sr_{1-x}M_xTiO_{3\pm d}$ perovskites in methane catalytic flameless combustion. J. Mol. Catal. A: Chem. **245**, 55–61 (2006)
56. Shannon, R.D.: Revised effective ionic radii and systematic studies of interatomic distances in halides and chalcogenides. Acta Crystallogr., Sect A **32**, 751–767 (1976)
57. Oliva, C., Bonoldi, L., Cappelli, S., Fabbrini, L., Rossetti, I., Forni, L.: Effect of preparation parameters on $SrTiO_3 \pm d$ catalyst for the flameless combustion of methane. J. Mol. Catal. A: Chem. **226**, 33–40 (2005)
58. Meriaudeau, P., Védrine, J.C.: Electron paramagnetic investigation of oxygen photoadsorption and its reactivity with carbon monoxide and titanium dioxide: the O_3^{3-} species. J. Chem. Soc., Faraday Trans. II **72**, 472–480 (1976)
59. Gonzalez-Elipe, A.R., Munuera, G., Soria, J.: Photo-adsorption and photo-desorption of oxygen on highly hydroxylated TiO_2 surfaces. J. Chem. Soc., Faraday Trans. I **75**, 748–761 (1979)
60. Takita, Y., Iwamoto, M., Lunsford, J.: Surface reactions of oxygen ions. 4. Oxidation of alkenes by O3—on magnesium oxide. J. Phys. Chem. **84**, 1710–1712 (1980)
61. Attwood, A.L., Murphy, D.M., Edwards, J.L., Egerton, T.A., Harrison, R.W.: An EPR study of thermally and photochemically generated oxygen radicals on hydrated and dehydrated titania surfaces. Res. Chem. Intermed. **29**, 449–465 (2003)

62. Howe, R.F., Timmer, W.C.: ESR Studies of O_2^- adsorbed on silica gel: photoformation and rotational dynamics. J. Chem. Phys. **85**, 6129–6135 (1986)
63. Dyrek, K., Adamski, A., Sojka, Z.: EPR study of the mobility of paramagnetic species on the surface and in the bulk of solids. Spectrochim. Acta A **54**, 2337–2348 (1998)
64. Bleaney, B.: Electron Spin resonance of paramagnetic impurities in antiferromagnetic compounds. Proc. R. Soc. Lond. A **433**, 461–468 (1991)
65. Shames, A.I., Rozenberg, E., Markovich, V., Auslender, M.I., Yakubovsky, A., Maignan, A., Martin, C., Raveau, B., Gorodetsky, G.: FMR probing of 'spontaneous' and Ru-doping induced ferromagnetism in $Sm_{0.2}Ca_{0.8}Mn_{1-x}Ru_xO_3$ ($x = 0.08$) manganites. Solid State Commun. **126**, 395–399 (2003)
66. Smit, J.: Mechanism for the direct exchange interaction of europium in compounds with rocksalt structure. J. Appl. Phys. **37**, 1455–1457 (1966)
67. Nagaev, E.L.: Physics of Magnetic Semiconductors. MIR, Moscow (1983)
68. Suzuki, E., Nakashiro, K.: Ono, Y.: Hydroxylation of Benzene with nitrogen monoxide over H-ZSM-5 Zeolite. Chem. Lett. 953–956 (1988)
69. Gubelmann, M.H., Tirel, P.J.: France Patent No. 2630735 (1988)
70. Kharitonov, A.S., Aleksandrova, T.N., Vostrikova, L.A., Sobolev, V.I., Ione, K.G., Panov, G.I.: USSR Patent No. 1805127 (1988)
71. Panov, G.I.: Advances in oxidation catalysis; oxidation of benzene to phenol by nitrous oxide. CATTech **4**, 18–31 (2000)
72. Ribera, A., Arends, I.W.C.E., de Vries, S., Perez-Ramirez, J., Sheldon, R.A.: Preparation, characterization, and performance of FeZSM-5 for the selective oxidation of benzene to phenol with N_2O. J. Catal. **195**, 287–297 (2000)
73. Kotasthane, A.N., Shiralkar, V.P., Hegde, S.G., Kulkarni, S.B.: Synthesis and characterization of alumino and ferrisilicate pentasil zeolites. Zeolites **6**, 253–260 (1986)
74. Volodin, A.M., Dubkov, K.A., Lund, A.: Direct ESR detection of S = 3/2 states for nitrosyl iron complexes in FeZSM-5 zeolites. Chem. Phys. Lett. **333**, 41–44 (2001)
75. Calis, G., Frenken, P., de Boer, E., Swolfs, A., Hefni, M.A.: Synthesis and spectroscopic studies of Fe^{3+} substituted ZSM-5 zeolite. Zeolites **7**, 319–326 (1987)
76. Kucherov, A.V., Slinkin, A.A.: Introduction Fe(III) ions in cationic positions of HZSM-5 by a solid-state reaction, Fe(III) cations in HZSM-5, and Fe(III) lattice ions in ferrisilicate. Zeolites **8**, 110–116 (1988)
77. Lin, D.H., Coudurier, G., Vedrine, J.C.: In: Jacobs, P.A., van Santen, R.A. (eds.) Zeolites: Facts, Figures, Future. Elsevier, Amsterdam (1989)
78. Bruckner, A., Luck, R., Wieker, W., Fahlke, B., Mehner, H.: E.p.r. study on the incorporation of Fe(III) ions in ZSM-5 zeolites in dependence on the preparation conditions. Zeolites **12**, 380–385 (1992)
79. Goldfarb, D., Bernardo, M., Strohmaier, K.G., Vaughan, D.E.W., Thomann, H.: Characterization of Iron in Zeolites by X-band and Q-Band ESR, Pulsed ESR, and UV-Visible Spectroscopies. J. Am. Chem. Soc. **116**, 6344–6353 (1994)
80. Bordiga, S., Buzzoni, R., Geobaldo, F., Lamberti, C., Giamello, E., Zecchina, A., Leofanti, G., Petrini, G., Tozzola, G., Vlaic, G.: Structure and reactivity of framework and extraframework iron in Fe-silicalite as investigated by spectroscopic and physicochemical methods. J. Catal. **158**, 486–501 (1996)
81. Weckhuysen, B.M., Wang, D.J., Rosynek, M.P., Lunsford, J.H.: Structure and reactivity of framework and extra framework Iron in Fe-silicalite as investigated by spectroscopic and physicochemical methods. Angew. Chem. Int. Ed. **36**, 2374–2376 (1997)
82. Brabec, L., Jeschke, M., Klik, R., Novakova, J., Kubelkova, L., Meusinger, J.J.: Appl. Catal. A: Fe in MFI metallosilicates, characterization and catalytic activity. General **170**, 105–116 (1998)
83. Kucherov, A.V., Kucherova, T.N., Slinkin, A.A.: Modification of zeolites by multi-charged cations by the use of in-situ formed "active gas-phase species". Microporous Mesoporous Mater. **26**, 1–10 (1998)

84. Kucherov, A.V., Montreuil, C.N., Kucherova, T.N., Shelef, M.: In situ high-temperature ESR characterization of FeZSM-5 and FeSAPO-34 catalysts in flowing mixtures of NO, C_3H_6, and O_2. Catal. Lett. **56**, 173–181 (1998)
85. Volodin, A.M., Sobolev, V.I., Zhidomirov, G.M.: The in situ ESR study of the state of irons in Fe-ZSM-5 zeolites. Kinet. Catal. **39**, 775–787 (1998)
86. El-Malki, E.M., van Santen, R.A., Sachtler, W.M.H.: Introduction of Zn, Ga, and Fe into HZSM-5 cavities by sublimation: identification of acid sites. J. Phys. Chem. **103**, 4611–4622 (1999)
87. Long, R.Q., Yang, R.T.: Characterization of Fe-ZSM-5 catalyst for selective catalytic reduction of nitric oxide by ammonia. J. Catal. **194**, 80–90 (2000)
88. El-Malki, E.M., van Santen, R.A., Sachtler, W.M.H.: Active sites in Fe/MFI catalysts for NOx reduction and oscillating N_2O decomposition. J. Catal. **196**, 212–223 (2000)
89. Fejes, P., Nagy, J.B., Lázar, K., Halász, J.: Heat-treatment of isomorphously substituted ZSM-5 (MFI) zeolites. An ESR and Mössbauer spectroscopy and kinetic study. Appl. Catal. A: General **190**, 117–135 (2000)
90. Chen, H.Y., El-Malki, E.M., Wang, X., van Santen, R.A., Sachtler, W.M.H.: Identification of active sites and adsorption complexes in Fe/MFI catalysts for NO_x reduction. J. Mol. Catal. A. Chem. **162**, 159–174 (2000)
91. Kucherov, A.V., Nissenbaum, V.D., Kucherova, T.N., Kustov, L.M.: Catalytic oxidation of methane by nitrous oxide on H[Al]ZSM-5 zeolite, silicalite, and amorphous SiO_2 modified by iron, silver, and gadolinium ions. Kinet. Catal. **43**, 711–723 (2002)
92. Kubanek, P., Wichterlova, B., Sobalik, Z.: Nature of active sites in the oxidation of benzene to phenol with N_2O over H-ZSM-5 with low Fe concentrations. J. Catal. **211**, 109–118 (2002)
93. Berlier, G., Spoto, G., Fisicaro, P., Bordiga, S., Zecchina, A., Giamello, E., Lamberti, C.: Co-ordination and oxidation changes undergone by iron species in Fe-silicalite upon template removal, activation and interaction with N_2O: an in situ X-ray absorption study. Microchem. J. **71**, 101–116 (2002)
94. Weil, J.A., Bolton, J.R., Wertz, J.E.: Electron Paramagnetic Resonance. Wiley, New York (1994)
95. Pilbrow, J.R.: Transition Ion electron paramagnetic resonance. Clarendon Press, Oxford (1990)
96. Ferretti, A.M., Oliva, C., Forni, L., Berlier, G., Zecchina, A., Lamberti, C.: Evolution of extraframework iron species in Fe-silicalite: 2. Effect of steaming. J. Catal. **208**, 83–88 (2002)
97. Le Caër, G., Brand, R. A.: General models for the distributions of electric field gradients in disordered solids. J. Phys.-Condens. Matter **10**, 10715–10774 (1998)
98. Ferretti, A.M., Barra, A.-L., Forni, L., Schweiger, A., Ponti, A.: Electron paramagnetic resonance spectroscopy of iron(III)-doped MFI zeolite. 1. multifrequency CW-EPR. J. Phys. Chem. B **108**, 1999–2005 (2004)
99. Ferretti, A.M., Forni, L., Oliva, C., Ponti, A.: EPR study of iron-doped MFI zeolite and silicalite catalysts: effect of treatments after synthesis. Res. Chem. Interm. **28**, 101–116 (2002)
100. Dorman, J.L., Fiorani, D., Tronc, E., Prigogine, I., Rice, S.A. (eds.): Magnetic relaxation in fine-particle systems. Adv. Chem. Phys. **98**, 283–494 (1997)
101. Ibrahim, M.M., Edwards, G., Seehra, M.S., Ganguly, B., Huffman, G.P.: Magnetism and spin dynamics of nanoscaie FeOOH particles. J. Appl. Phys. **75**, 5873–5875 (1994)
102. Hagiwara, M., Nagata, K., Nagata, K.: Magnetism and magnetic interaction in a complex oxide glass system containing deposited clusters of magnetite at the superparamagnetic state. J. Phys. Soc. Japan **67**, 3590–3600 (1998)
103. Berger, R., Kliava, J., Bissey, J.C.: Magnetic resonance of superparamagnetic iron-containing nanoparticles in annealed glass. J. Appl. Phys. **87**, 7389–7396 (2000)
104. Oliva, C., Forni, L.: EPR investigation of superparamagnetism in nanometric-size LaFeO3 $\pm$ d and La2NiO4 $\pm$ d. Perovskites. Appl. Magn. Reson. **20**, 531–538 (2001)

105. Sharma, V.K., Wakdner, F.: Superparamagnetic and ferrimagnetic resonance of ultrafine Fe_30_4 particles in ferrofluids. J. Appl. Phys. **48**, 4298–4301 (1977)
106. Cannas, C., Gatteschi, D., Musinu, A., Piccaluga, G., Sangregorio, C.: Structural and magnetic properties of Fe_2O_3 nanoparticles dispersed over a silica matrix. J. Phys. Chem. **102**, 7721–7726 (1998)
107. Berger, R., Kliava, J., Bissey, J.C., Baietto, V.: Magnetic resonance of superparamagnetic iron-containing nanoparticles in annealed glass. J. Appl. Phys. **87**, 7389–7396 (2000)
108. A.I.P. Hanbook, Sect.5g, 1963; Landolt-Bornstein, vol. 2, pt9, 6th ed. (1962)
109. Meloni, D., Monaci, R., Solinas, V., Berlier, G., Bordiga, S., Rossetti, I., Oliva, C., Forni, L.: Activity and deactivation of Fe-MFI catalysts for benzene hydroxylation to phenol by N_2O. J. Catal. **214**, 169–178 (2003)
110. Bonneviot, L., Olivier, D.: Ferromagnetic resonance. In: Imelik, B., Védrine, J.C. (eds.) Catalyst characterization. Physical Techniques for Solid Materials (1994)
111. Sharma, V.K., Waldner, F.: Superparamagnetic and ferrimagnetic resonance of ultrafine Fe_3O_4 particles in ferrofluids. J. Appl. Phys. **48**, 4298–4302 (1977)
112. Raikher, Y.L., Stepanov, V.I.: Ferromagnetic resonance in a suspension of single-domain particles. Phys. Rev. B **50**, 6250–6259 (1994)
113. Berger, R., Kliava, J., Bissey, J.-C., Baietto, V.: Magnetic resonance of superparamagnetic iron-containing nanoparticles in annealed glass. J. Appl. Phys. **87**, 7389–7396 (2000)
114. Koksharov, Yu.A, Pankratov, D.A., Gubin, S.P., Kosobudsky, I.D., Beltran, M., Khodorkovsky, Y., Tishin, A.M.: Electron paramagnetic resonance of ferrite nanoparticles. J. Appl. Phys. **89**, 2293–2319 (2001)
115. Abragam, A., Bleaney, B.: Electron Paramagnetic Resonance of transition ions, pp. 308–312. Dover Publications, New York (1986)
116. Abragam, A., Bleaney, B.: Electron Paramagnetic Resonance of Transition Ions, Chapter 18. Dover Publications, New York (1986)
117. Bagguley, D.M.S., Vella-Coleiro, G.: Pulsed field Zeeman studies of Ce^{3+} in $LaCl_3$. J. Phys. C: Solid State Phys. **2**, 2310–2319 (1969)
118. Birgenau, R.J.: Magnetic moment reduction via the orbit-lattice interaction for Ce^{3+} in rare-earth ethyl sulfates. Phys. Rev. Lett. **19**, 160–162 (1967)
119. Judd, B.R.: The magnetic and spectroscopic properties of certain rare-earth double nitrates. Proc. R. Soc., A **232**, 458–474 (1955)
120. Kreissl, J.: Electron paramagnetic resonance of Ce^{3+} in SrS powder. Phys. Stat. Sol. (b), **180**, 441–443 (1993)
121. Kindo, K., Shibata, T., Inoue, T., Haga, Y., Suzuki, T., Chiba, Y., Date, M.: Oscillatory high-field magnetization in LaP doped with Ce. J. Phys. Soc. Japan. **62**, 4190–4193 (1993)
122. Inaba, H., Tagawa, H.: Ceria-based solid electrolytes. Sol. State Ionics **83**, 1–16 (1996)
123. Steel, B.C.H.: In: Tuller, H.L., Schoonman, J., Riess, I. (eds.) Oxygen Ion and Mixed Conductors and their Technological Applications. NATO ASI Series, Serie E, Applied Sciences, vol. 368, pp. 323–345 (1997)
124. Hwang, J.-H., Mason, O.: Defect chemistry and transport properties of nanocrystalline cerium oxide. Z. Physikalische Chemie. **207**, 21–38 (1998)
125. Chiang, Y.-M., Lavik, E.B., Kosacki, I., Tuller, H.L., Ying, J.Y.: Defect and transport properties of nanocrystalline CeO_{2-x}. Appl. Phys. Lett. **69**, 185–187 (1996)
126. Oliva, C., Scavini, M., Ballabio, O., Sin, A., Zaopo, A., Dubitsky, Y.: Percolative Small-polarons conduction regime in $Ce_{1-x}Gd_xO_{(4-x)/2}$ probed by EPR spectral intensity of Gd^{3+}. J. Solid State Chem. **177**, 4104–4111 (2004)
127. Bonneviot, L., Olivier, D.: Ferromagnetic resonance. In: Imelik B., Védrine, J.C. (eds.) Catalyst Characterization. Physical Techniques for Solid Materials (1994)
128. Guskos, N., Likodimos, V., Typek, J., Wabia, M., Fuks, H.: EPR spectra of Gd^{3+} ions and localized paramagnetic centers in $La_{0.5}Gd_{0.5}Ba_3Cu_2O_{6+?}$. Physica C **341–348**, 573–574 (2000)
129. Guskos, N., Triberis, G.P., Lykodimos, V., Windsch, W., Metz, H., Koufoudakis, A., Mitros, C., Gamari-Seale, H., Niarchos, D.: Temperature dependence of the EPR spectra of

$Gd_{0.5}RE_{0.5}$ $Ba_2Cu_3O_{7-d}$ compounds in the orthorhombic and tetragonal phases. Phys. Stat. Sol. (b) **166**, 233–244 (1991)
130. Guskos, N., Kuriata, J., Salikhov, I.H.: The anomalous electric field effect in the EPR of CdF_2: Gd^{3+}. J. Phys. C: Solid State Phys. **17**, 2175–2180 (1984)
131. Triberis, G.P., Friedman, L.R.: A percolation treatment of the conductivity for the high temperature small-polaron hopping regime in disordered systems. J. Phys. C: Solid State Phys. **14**, 4631–4639 (1981)
132. Ohashi, T., Yamazaki, S., Tokunaga, T., Arita, Y., Matsui, T., Harami, T., Kobayashi, K.: EXAFS study of $Ce_{1-x}Gd_xO_{2-x/2}$. Solid State Ionics **113–115**, 559–564 (1998)
133. Nakagawa, T., Osuki, T., Yamamoto, T.A., Kitauji, Y., Kano, M., Katsura, M., Emura, S.: Study on local structure around Ce and Gd atoms in CeO_2–Gd_2O_3 binary system. J. Synchrotron Rad. **8**, 740–742 (2001)
134. Oliva, C., Orsini, F., Cappelli, S., Arosio, P., Allieta, M., Coduri, M., Scavini, M.: Electron spin resonance and atomic force microscopy study on gadolinium doped ceria. J. Spectrosc. **491840**, 1–6 (2015). doi:10.1155/2015/491840
135. Scavini, M., Coduri, M., Allieta, M., Masala, P., Cappelli, S., Oliva, C., Brunelli, M., Orsini, F., Ferrero, C.: Percolation hierarchical defect structures drive phase transformation in $Ce_{1-x}Gd_xO_{2-x/2}$: a total scattering study. IUCrJ **2**, 511–522 (2015)
136. Oliva, C., Allieta, M., Scavini, M., Biffi, C., Rossetti, I., Forni, L.: EPR analysis of La1-xMxMnO3 + d (M = Ce, Sr) perovskite-like nanostructured catalysts. Inorg. Chem. **51**, 8433–8440 (2012)
137. Allieta, M., Scavini, M., Lo Presti, L., Coduri, M., Loconte, L., Cappelli, S., Oliva, C., Ghigna P., Pattison, P., Scagnoli, V.: Charge ordering transition in $GdBaCo_2O_5$: Evidence of reentrant behavior. Phys. Rev. B **88**, 1–10 (2013)
138. Allieta, M., Oliva, C., Scavini, M., Cappelli, S., Pomjakushina, E., Scagnoli, V.: The nature of spin-lattice interaction in the insulator to metal transition of GdBaCo2O5 + d. Phys. Rev. B **84**, 1–7 (2011)
139. Sugawara, K., Huang, C.Y.: Electron paramagnetic resonance of Gd^{3+} in Yttrium-Group VA intermetallic compounds. J. Phys. Soc. Japan **39**, 643–646 (1975)
140. Bejjitt, L., Haddad, M.: Oxygen content dependence of the EPR line shape in $GdBa_2Cu_30_{7-y}$ single crystal. Appl. Magn. Reson. **22**, 415–419 (2002)
141. Urbano, R.R., Pagliuso, P.G., Rettori, C., Schlottmann, P., Fisk, Z., Chapler, B., Oseroff, S. B.: Gd $^{3+}$ and Eu $^{2+}$ local environment in $Ca_{1-x}Eu_xB_6$ (0.0001 = x = 0.30) and $Ca_{1-x}Gd_xB_6$ (0.0001 = x = 0.01). Phys. Status Solidi (a) **203**, 1550–1555 (2006)
142. Oliva, C., Scavini, M., Ballabio, O., Sin, A., Zaopo, A., Dubitsky, Yu.: Percolative small-polarons conduction regime in $Ce_{1-x}Gd_xO_{(4-x)/2}$ probed by EPR spectral intensity of Gd^{3+}. J. Solid State Chem. **177**, 4104–4111 (2004)
143. Kaplan, J.I., Reuben, J.: Electron spin resonance line shapes of paramagnetic species on surfaces. J. Phys. Chem. **86**, 4465–4466 (1982)
144. Skakle, M.S.: Crystal chemical substitutions and doping of YBa_2Cu_3Ox and related superconductors. Mater. Sci. Eng. R **23**, 1–40 (1998). and references therein
145. Tolentino, H., Baudelet, F., Fontaine, A., Gourieux, T., Krill, G., Henry, J.Y., Rossat-Mignod, J.: Sequence and symmetry of hole injection in $YBa_2Cu_3O_{6+x}$ in situ and extra situ experiments on powders and single crystals using X-ray absorption spectroscopy. Physica C **192**, 115–130 (1992)
146. Guillaume, M., Allenspach, P., Henggeler, W., Mesot, J., Roessli, B., Staub, U., Fischer, P., Furrer, A., Trounov, V.: A systematic low-temperature neutron diffraction study of the $RBa_zCu_30_x$ (R = yttrium and rare earths; x: = 6 and 7) compounds. J. Phys.: Condens. Matter **6**, 7963–7976 (1994)
147. Licci, F., Gauzzi, A., Marezio, M., Radaelli, G.P., Masini, R., Chatillout-Boougerol, C.: Structural and electronic effects of Sr substitution for Ba in $Y(Ba_{1-x}Sr_x)_2Cu_3O_w$ at varying w. Phys. Rev. B **58**, 15208–15217 (1998)
148. Scavini, M., Bianchi, R.: Structure of Al defect in high-temperature superconductor, Al-doped Sm-123: an electron density study. J. Sol. State. Chem. **161**, 396–401 (2001)

149. Scavini, M., Daldosso, M., Cappelli, S., Oliva, C., Brunelli, M., Ferrero, C., Lascialfari, A.: Unravelling the role of Cu-O chain length on the superconducting properties of SmBa2Cu3O6 + d via Al doping and clustering. Europhys. Lett. **76**, 443–449 (2006)
150. Dupree, R., Gencten, A., Paul, D.M.: A ^{89}Y NMR study of substitution for copper in $YBa_2(Cu_{1-x}M_x)_3O_7$. Physica C **193**, 81–89 (1992)
151. Guo, Y., Langlois, J.-M., Goddard III, W.A.: Electronic structure and valence-bond band structure of cuprate superconducting materials. Science **239**, 896–899 (1988)
152. Likodimos, V., Guskos, N., Gamari-Seale, H., Koufoudakis, A., Wabia, M., J.Typek, J, Fuks, H.: Copper magnetic centers in oxygen deficient $RBa_2Cu_3O_{6+x}$ (R = Nd, Sm): an EPR and magnetic study. Phys. Rev. B. **54**, 12342–12352 (1996)
153. Eremin, M.V., Sigmund, F.: On the importance of anisotropic exchange coupling on to the structure of spin-polarized clusters in layered cuprates. Solid State Commun. **90**, 795–797 (1994)
154. Thomann, H., Klemm, R.A., Johnston, D.C., Tindall, P.J., Goshorn, D.P., Jin, H.: Observation of triplet hole pairs and glassy spin waves in $La_{2-x-z}Sr_xCuO_{4-y}$ by electron spin resonance. Phys. Rev. B **38**, 6552–6560 (1988)
155. Forni, L., Oliva, C., Vatti, F.P., Kandala, M.A., Ezerets, A.M., Vishniakov, A.V.: La-Ce-Co perovskites as catalysts for exhaust gas depollution. Appl. Catal. B: Environ. **7**, 269–284 (1996)
156. Forni, L., Oliva, C., Barzetti, T., Selli, E., Ezerets, A.M., Vishniakov, A.V.: FT-IR and EPR spectroscopic analysis of $La_{1-x}Ce_xCoO_3$ perovskite-like catalysts for NO reduction by CO. Appl. Catal. B: Environ. **13**, 35–43 (1997)
157. Leanza, R., Rossetti, I., Fabbrini, L., Oliva, C., L.Forni, L.: Perovskite catalysts for the catalytic flameless combustion of methane Preparation by flame-hydrolysis and characterisation by TPD-TPR-MS and EPR Appl. Catal. B: Envirnon. **28**, 55–64 (2000)
158. Oliva, C., Forni, L., D'Ambrosio, A., Navarrini, F., Stepanov, A.D., Kagramanov, Z.D., Mikhailichenko, A.I.: Characterisation by EPR and other techniques of La1-xCexCoO3 + d perovskite-like catalysts for methane flameless combustion. Appl. Catal. A: General **205**, 245–252 (2001)
159. Abragam, A., Bleaney, B.: Electron Paramagnetic Resonance of Transition Ions, Chaps. 9 and 10. Dover, New York (1986)
160. Oliva, C., Forni, L., Formaro, L.: Effect of thermal treatment on the EPR spectra and on catalytic properties of pure Co_3O_4. Appl. Spectrosc. **50**, 1395–1398 (1996)
161. Oliva, C., Cappelli, S., Kryukov, A., Chiarello, G.L., Vishniakov, A.V., Forni, L.: Effect of preparation parameters on the properties of $La_{0.9}Ce_{0.1}CoO_3$ catalysts: an EMR investigation. J. Mol. Catal. A Chem. **255**, 36–40 (2006)
162. Chiarello, G.L., Rossetti, I., Forni, L.: Flame-spray pyrolysis preparation of perovskites for methane catalytic combustion. J. Catal. **236**, 251–261 (2005)
163. Angelov, S., Zhecheva, E., Stoyanova, R., Atanasov, M.: Bulk defects in Co_3O_4, pure and slightly doped with lithium, revealed by EPR of the tetrahedral Co^{2+} ions. J. Phys. Chem. Solids **51**, 1157–1161 (1990)
164. Roth, W.L.: The magnetic structure of Co_3O_4. J. Phys. Chem. Solids **25**, 1–10 (1964)
165. Tanaka, H., Mizuno, N., Misono, M.: Catalytic activity and structural stability of $La_{0.9}Ce_{0.1}Co_{1-x}Fe_xO_3$ perovskite catalysts for automotive emissions control. Appl. Catal. A: General **244**, 371–382 (2003)
166. Choi, Y.M., Lynch, M.E., Lin, M.C., Liu, M.: Prediction of O_2 dissociation kinetics on $LaMnO_3$-based cathode materials for solid oxide fuel cells. J. Phys. Chem. C **113**, 7290–7297 (2009)
167. Kaddouri, A., Gelin, P., Dupont, N.: Methane catalytic combustion over La–Ce–Mn–O–perovskite prepared using dielectric heating. Catal. Commun. **10**, 1085–1089 (2009)
168. Li, Y., Xue, L., Fan, L., Yan, Y.: The effect of citric acid to metal nitrates molar ratio on sol–gel combustion synthesis of nanocrystalline $LaMnO_3$ powders. J. Alloys Compd. **478**, 493–497 (2009)

169. Rossetti, I., Buchneva, O., Biffi, C., Rizza, R.: Effect of sulphur poisoning on perovskite catalysts prepared by flame-pyrolysis. Appl. Catal. B: Environ. **89**, 383–390 (2009)
170. Peña, M.A., Fierro, J.L.G.: Chemical structures and performance of perovskite oxides. Chem. Rev. **101**, 1981–2018 (2001)
171. Tsai, Y.T., Chang, W.J., Wuang, S.W., Lin, J.-Y., Lee, J.Y., Chen, J.M., Wu, K.H., Uen, T. M., Gou, Y.S., Juang, J.Y.: Temperature-dependent X-ray absorption near edge spectroscopy of doped LaMnO3: ion-size and the Jahn-Teller distortion effects. Phys. B **404**, 1404–1408 (2009)
172. Oliva, C., Forni, L., Pasqualin, P., D'Ambrosio, A., Vishniakov, A.V.: EPR analysis of La_{1-x} M_x MnO_{3+y} (M = Ce, Eu, Sr) perovskitic catalysis for methane oxidation. Phys. Chem. Chem. Phys. **1**, 355–360 (1999)
173. Ivanshin, V.A., Deisenhofer, J., Krug von Nidda, H.A., Loidl, A., Mukhin, A.A., Balbashov, A.M., Eremin, M.V.: ESR study in light doped $La_{1-x}Sr_xMnO_3$. Phys. Rev. B **61**, 6213–6219 (2000)
174. Noginova, N., Bah, R., Bitok, D., Atsarkin, V.A., Demidov, V.V., Gudenko, S.V.: Effect of diamagnetic dilution and non-stoichiometry on ESR spectra of manganites. J. Phys.: Condens. Matter **17**, 1259–1269 (2005)
175. Mitchell, J.F., Argyriou, D.N., Potter, C.D., Hinks, D.G., Jorgensen, J.D., Bader, S.D.: Structural phase diagram of $La_{1-x}Sr_xMnO_{3+d}$: relationship to magnetic and transport properties. Phys. Rev. B **54**, 6172–6183 (1996)
176. Rodrìguez-Carvajal, J., Hennion, M., Moussa, F., Moudden, A.H., Pinsard, L., Revcolevschi, A.: Neutron-diffraction study of the Jahn-Teller transition in stoichiometric $LaMnO_3$. Phys. Rev. B **57**, R3189–R3192 (1998)
177. Oliva, C., Forni, L.: EPR and XRD as probes for activity and durability of $LaMnO_3$ perovskite-like catalysts. Catal. Commun. **1**, 5–8 (2000)
178. Strobel, R., Pratsinis, S.E.: Flame aerosol synthesis of smart nanostructured materials. J. Mater. Chem. **17**, 4743–4756 (2007)
179. Chiarello, G.L., Rossetti, I., Forni, L., Lopinto, P., Migliavacca, G.: Solvent nature effect in preparation of perovskites by flame-pyrolysis 1. Carboxylic acids. Appl. Catal. B: Environ. **72**, 218–226 (2007) and: Solvent nature effect in preparation of perovskites by flame-pyrolysis 2. Alcohols and alcohols + propionic acid mixtures. Appl. Catal. B: Environm. **72**, 227–232 (2007)
180. Feher, G., Kip, A.F.: Electron spin resonance absorption in metals. I. Exp. Phys. Rev. **98**, 337–348 (1955)
181. Dyson, F.J.: Electron spin resonance absorption in metals. II. Theory of electron diffusion and the skin effect. Phys. Rev. **98**, 349–359 (1955)
182. Moskvin, A.S.: Pseudo-Jahn-Teller-centers and phase separation in the strongly correlated oxides with the nonisovalent substitution. Cuprates and manganites. Physica B **252**, 186–197 (1998)
183. Moskvin, A.S., Avvakumov, I.D.: Doped manganites beyond conventional double-exchange model. Phys. B **322**, 371–389 (2002)
184. Moskvin, A.S.: Disproportionation and electronic phase separation in parent manganite $LaMnO_3$. Phys. Rev. B **79**, 115102-1-19 (2009)
185. Asamitsu, A., Moritomo, Y., Tomioka, Y., Arima, T., Tokura, Y.: A structural phase transition induced by an external magnetic field. Nature **373**, 407–409 (1995)

EPR Studies of Cerium Dioxide Nanoparticles

Ashutosh Kumar Shukla and Rafail Rakhmatullin

Abstract This chapter intends to provide a review of EPR spectroscopic characterization of ceria nanoparticles. It covers recent EPR applications to investigate the synthesis, local structure, and distortion of ceria nanoparticles. Room temperature ferromagnetism, toxicological screening, free radical scavenging properties, catalytic effects, and some other properties of ceria nanoparticles as studied by EPR are also included in this chapter.

1 Introduction

Cerium dioxideor ceria (CeO_2) nanoparticles have found energy related technological applications such as solar cells and solid oxide fuel cells [1]. It has attracted the attention due to its wide band gap and high dielectric constant [2]. Increasing the catalytic activity of nanocrystalline ceria for applications in diesel particulate filters for removal of soot from diesel and organic compounds from wastewater has also attracted the researchers [3]. Ceria nanoparticles also have antioxidant properties. Their antioxidative ability lies in the ability of Ce ions to switch between different valence states, i.e., from Ce^{3+} to Ce^{4+}. This has opened up new dimensions for application of ceria nanoparticles against radiation damage, oxidative stress, and inflammation [4]. Toxicological screening of ceria nanoparticles has also gained importance due to expected occupational exposure in the wide range of applications. Enhancing the photoactivity of CeO_2 in the visible range is another area of interest which has led to the modifications like doping in CeO_2. Electron Paramagnetic Resonance spectroscopy has suitably been applied in the study of undoped and doped ceria nanoparticles. The application of CW and pulsed EPR

A.K. Shukla (✉)
Physics Department, Ewing Christian College, Allahabad 211003, UP, India
e-mail: drakshukla@gmail.com

R. Rakhmatullin
Institute of Physics, Kazan Federal University, Russian Federation, 420008 Kazan, Russia

A.K. Shukla (ed.), *EMR/ESR/EPR Spectroscopy for Characterization of Nanomaterials*, Advanced Structured Materials 62,
DOI 10.1007/978-81-322-3655-9_6

techniques can provide new information that may lead to improvement in the performance of ceria-based devices. We have therefore made an attempt to review recent EPR studies of undoped and doped ceria nanoparticles in this chapter.

2 Synthesis, Local Structure, and Distortions

Araujo et al. (2012) have reported the synthesis of CeO_2 nanoparticles by a microwave-assisted hydrothermal method at different synthesis temperatures [5]. These nanoparticles were characterized by EPR spectroscopy. Figure 1 presents the EPR spectra of the samples. Authors have reported that pure CeO_2 revealed formation of one type of paramagnetic species, A, due to Ce^{3+}. Assumption of the presence of Ce^{3+} ions and oxygen vacancies in the samples could be confirmed.

To improve the catalytic and oxygen storage properties of ceria, it is important to know the distribution of oxygen vacancies and the crystal field change depending on size of nanoparticles and dopant concentration. Such kind of information can be obtained with the EPR technique using probe ions introduced in small concentration to ceria in the process of synthesis.

EPR of Gd^{3+} probe ions, to study yttrium doped ceria (YDC) with the formula $Ce_{1-x-y}Gd_xY_yO_{2-[0.5\ (x+y)]}$ ($x = 0.0025$, $y = 0.10$ and 0.25) with crystallite sizes ranging from 600 nm down to 5 nm, is reported by Rakhmatullin et al. [6]. All YDC samples were synthesized via a co-precipitation method, where an aqueous solution of NH_4OH was added drop wise to an aqueous solution containing the stoichiometric amounts of metal nitrates. The precipitated oxides were collected by centrifugation and were washed with water; a 50 % vol. ethanol–water solution and subsequently with pure ethanol. The resulting nanopowders were then dried at 120 °C for 12 h, ground and finally annealed under air at temperatures ranging between 450 and 650 °C for 1–2 h to obtain crystallites ranging between ~5 and 10 nm in diameter. Samples with larger grain size (600 ± 50 nm in diameter) were obtained from the nanocrystalline powders that were pelletized

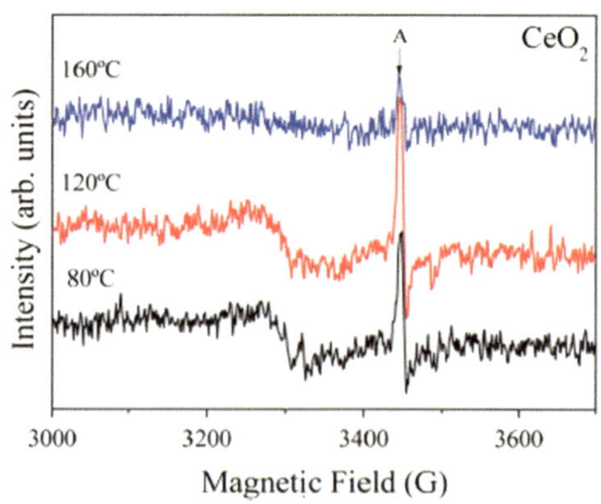

Fig. 1 EPR spectra of CeO_2 samples synthesized at different temperature. A represents the formation of paramagnetic species due to Ce^{3+}. Reproduced from Ref. [5] with permission of The Royal Society of Chemistry

via cold isostatic pressing at 276 MPa and subsequently sintered at 1400 °C for 12 h. X-ray diffraction (XRD) measurements of powder samples indicated that all micro and nanocrystalline doped ceria samples studied may be characterized by the cubic fluorite crystal structure. The crystallite size of the nanocrystalline samples could be estimated from widths of the four strongest peaks in the XRD patterns using the Williamson–Hall analysis. The particle size of the microcrystalline samples was estimated from field emission scanning electron microscopy (SEM) images. The Y contents were analyzed using energy dispersive x-ray spectroscopy in an SEM and were found to be within ± 1 cation% of the nominal composition in all cases. The unit cell dimension calculated from the XRD peak positions was found to be independent of the crystallite size (within ±0.005 Å) and decreased by ~0.01 Å with increasing Y content from ~10 to 25 cation%.

X-band CW EPR spectra of heavily Y-doped ceria samples were found to be similar to those observed in oxide glasses rather than those characteristic of Gd-doped CeO_2. Q-band spectra are dominated by the central transition that is located near about 12000 Gauss. X-band spectral line shapes were consistent with the expected low symmetry of the crystal field around Gd^{3+} ions in YDC samples, where the first coordination sphere of Gd would contain oxygen vacancies and a fraction of the 12 Ce^{4+} next nearest neighbors is substituted by Y^{3+} ions. In both X and Q-band spectra, the width of the central transition $|1/2\rangle \leftrightarrow |-1/2\rangle$ near $g \sim 2.0$ is found to decrease systematically with decreasing size of the crystal (Figs. 2 and 3) as well as with decreasing Y content (Fig. 4) and with increasing the temperature of measurement from ~6 K to room temperature.

The broadening of the central transition results from a distribution of crystal field parameters and in the case of a powder samples, also from a random distribution of crystallite orientations relative to the applied field. The approximate estimate of line width $\Delta\nu$ from perturbation theory for both types of scatter is given by $\Delta\nu \sim |H_{CF}|^2/h\nu$, where H_{CF} is the crystal field potential. The observation in the EPR spectra of progressive narrowing of the width of the central transition with decreasing crystallite size (Figs. 3 and 4) can be explained by a concomitant weakening of the crystal field around Gd^{3+} ions in the yttrium doped ceria lattice. This hypothesis is consistent with the observed narrowing of the width of the central transition with decreasing Y content and with increasing the temperature of measurement from ~6 K to room temperature. Both increasing temperature and decreasing Y content are known to result in lattice expansion of yttrium doped ceria, the former due to thermal expansion and the latter due to removal of oxygen vacancies [7, 8]. Such expansion of the lattice is expected to weaken the crystal field around Gd^{3+} ions in the YDC lattice resulting in a narrowing of the width of the central transition. [Reproduced with permission from J. Chem. Phys. **131**, 124515 (2009). Copyright 2009, AIP Publishing].

Authors could conclude that weakening of the average crystal field and narrowing of the EPR line width result from a more random oxygen vacancy distribution in the nanocrystals compared to their microcrystalline counterparts of same composition.

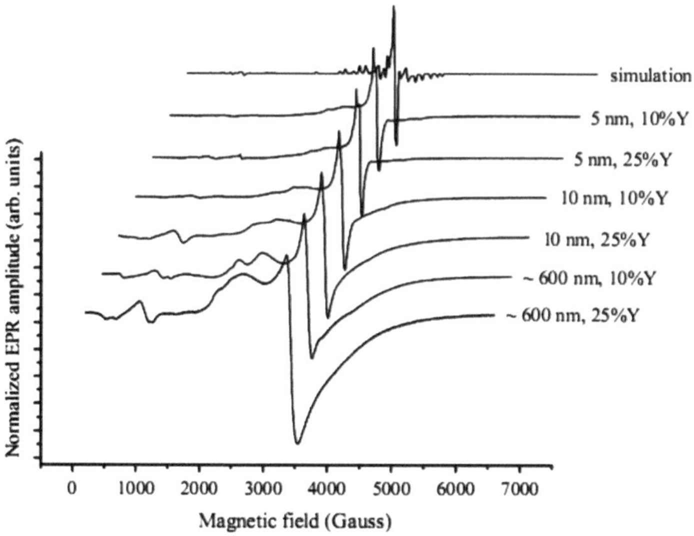

Fig. 2 CW EPR spectra of micro and nanocrystals of 10 and 25 cation% Y-doped ceria samples (co-doped with 0.25 % Gd) at X-band and liquid helium temperature (~6 K).The scale on the abscissa corresponds to the bottom-most spectrum. The other spectra are shifted horizontally for visual clarity. The topmost spectrum shows the simulation of the powder spectrum of Gd^{3+} in Y-free CeO_2. Reprinted with permission from J. Chem. Phys. **131**, 124515 (2009). Copyright 2009, AIP Publishing LLC

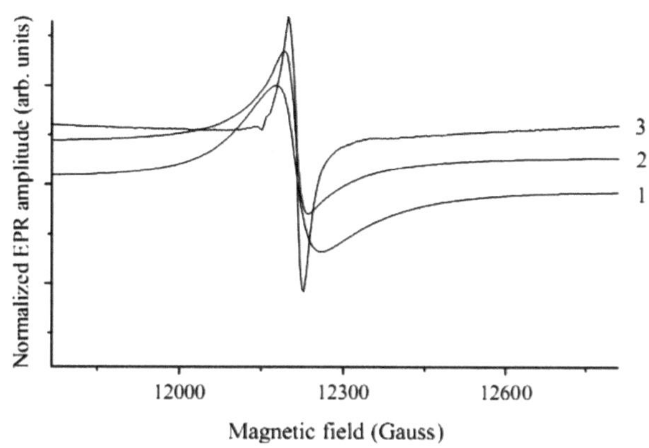

Fig. 3 The line width of (−½ ↔ +½ transition) as a function of crystallite size of the sample with 25 cation% Y at Q-band and liquid helium temperature (6 K). The labels 1, 2, and 3 correspond to crystallite sizes of ~600, 10, and 5 nm, respectively. Reprinted with permission from J. Chem. Phys. **131**, 124515 (2009). Copyright 2009, AIP Publishing LLC

EPR spectroscopy of Gd^{3+} probe ions to investigate size dependent lattice distortions in bulk and nanocrystalline CeO_2 with crystallite sizes of 600 (bulk) and 10 (nano) nm, respectively, and doped with 0.5 and 1 cation% Y is reported by Rakhmatullin et al. [9]. The studied samples were synthesized in the same way as earlier [6], except lower content of yttrium. It was found that incorporation of Gd^{3+} ions at 0.25 % cation concentration level does not introduce by itself any

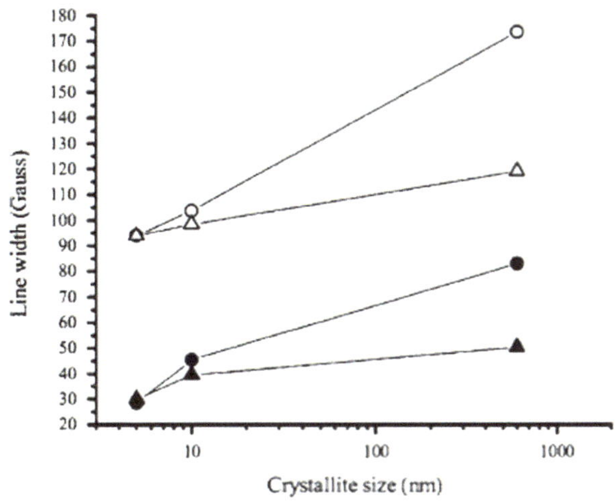

Fig. 4 The line width dependence of (−½ ↔ +½ transition) on the crystallite size of the sample at X (open symbols) and Q-band (filled symbols) and at liquid helium temperature (6 K). Circles and triangles correspond to the samples with 25 and 10 cation% Y, respectively. Reprinted with permission from J. Chem. Phys. **131**, 124515 (2009). Copyright 2009, AIP Publishing LLC

measurable strain in the CeO_2 lattice. EPR spectrum of the bulk sample with 0.5 cation% Y with simulation parameters of monocrystalline CeO_2 (Fig. 5) indicates the absence of any other significant contribution to the EPR line shape and suggests a nearly perfect cubic structure of the lattice of this sample.

Introduction of 1.0 cation% Y in bulk ceria lead more vacancies and clear distortions appear in the experimental spectrum. The simulation of the spectrum (Fig. 6) then yields the result that ~83 % of the Gd^{3+} sites have undistorted cubic symmetry and ~17 % of the Gd^{3+} sites are axially distorted from the presence of a neighboring oxygen vacancy.

The nanocrystalline samples display a different line shape, where the fine structure of Gd^{3+} is substantially broadened (Fig. 7). The simulations of EPR spectra for both nanocrystalline samples reveal the presence of two types of axial sites with rather different distortions that belong to the core and the surface regions of the nanocrystals and solely originate from crystallite size effect as the ratio of surface: bulk energy rapidly increases with decreasing size in nanocrystals. The

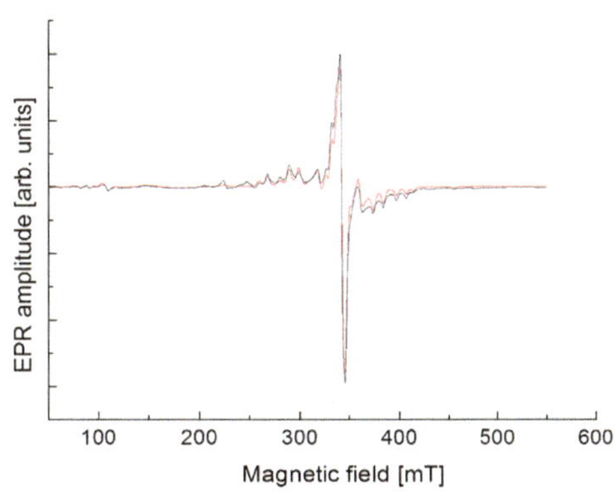

Fig. 5 The experimental (black line) and simulated (red line) EPR spectra of Gd^{3+} ions in the "bulk" ceria sample co-doped with 0.5 cation% Y. Reprinted with permission from [J. Appl. Phys. **114**, 203507 (2013)]. Copyright 2013, AIP Publishing LLC

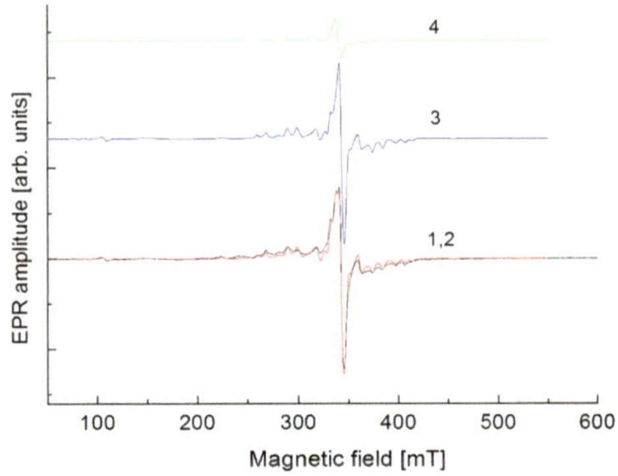

Fig. 6 Curves 1 and 2 represent experimental (*black*) and simulated (*red*) EPR spectra of Gd^{3+} ions in the "bulk" ceria sample co-doped with 1.0 cation% Y. Curve 3 (*blue*) represents simulated contribution from cubic sites, Curve 4 (*green*) represents simulated contribution from the trigonal sites. Reprinted with permission from [J. Appl. Phys. **114**, 203507 (2013)]. Copyright 2013, AIP Publishing LLC

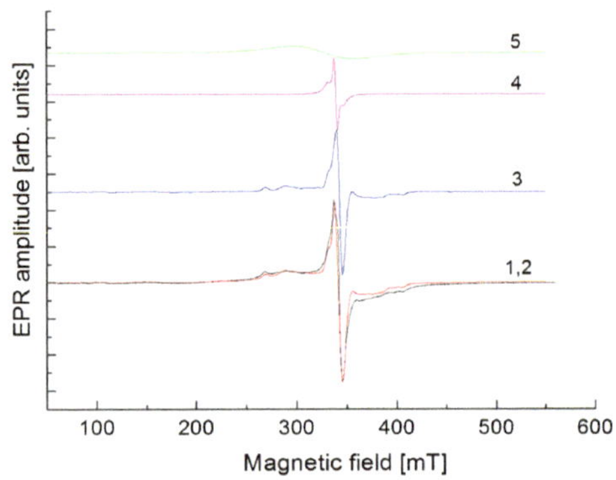

Fig. 7 Curves 1 and 2 show experimental (*black*) and simulated (red) EPR spectra of Gd^{3+} ions in the nanocrystalline ceria sample co-doped with 0.5 cation% Y, respectively. Curve 3 (*blue*) shows simulated contribution from the core trigonal sites, Curve 4 (*pink*) shows simulated contribution from the surface trigonal sites, Curve 5 (*green*) shows simulated contribution from surface superoxide species. Reprinted with permission from [J. Appl. Phys. **114**, 203507 (2013)]. Copyright 2013, AIP Publishing LLC

authors of this publication supposed that the size dependent lattice distortions may be universal in nanomaterials.

EPR studies of Er^{3+} probe ions in ceria nanoparticles with grain sizes of about 22 and 300 nm are reported [10, 11]. Ceria samples contained nominal concentration of 0.01 % of Er^{3+} ions in initial components. The nanocrystals synthesized via a co-precipitation method had crystallite diameter of 22 $\pm$ 2 nm (Sample 1) and 300 $\pm$ 10 nm (Sample 2). The purpose of these studies is to find out the vacancy distributions and the origin of unusual EPR lines with g-values around 14 and 20

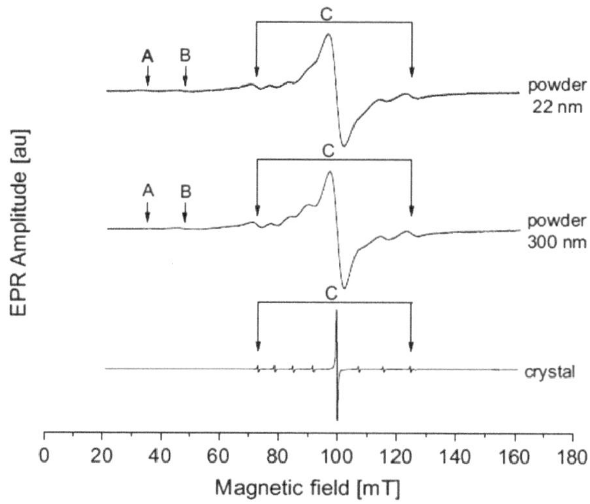

Fig. 8 X-band EPR spectra at 10 K of Er-doped CeO_2 single crystal (*bottom*) and nanocrystalline powder samples 1 (*top*) and 2 (*middle*) with average crystallite diameters of ~22 nm and ~300 nm, respectively. The line marked by C represents the cubic center with g = 6.76 along with arrows marking the spread of hyperfine lines from the odd isotope Er^{167}. The weak lines marked by A and B in the spectra of samples 1 and 2 are located near g ~20 and 14, respectively. [Reproduced with permission from Appl. Magn. Reson. **46**, 741 (2015)]

detected in nanocrystalline CeO_2.These lines are marked by A and B in the Fig. 8. The most intensive EPR line of Fig. 8, marked by "C" corresponds to the cubic site of Er^{3+} ions with g ≈6.75 due to the even isotopes and eight relatively weak lines correspond to the hyperfine splitting from the odd isotope of Er^{3+} (Er^{167}abundance is 22.9 %) with nuclear spin I = 7/2. The line width values ΔB_{pp} for the cubic center were 5.4, 4.8, and 0.2 mT for sample 1, sample 2 and single crystal, respectively. The larger widths of the EPR lines in the nanocrystals compared to those in the single crystal assume larger degree of local structural disorder in the former that increases with decreasing crystallite size. The concentration of the Er cubic centers as estimated from the EPR spectra was ~0.0050(5) at.% for sample 2 and ~0.0020(2) at.% for the sample 1. These values are smaller than the nominal concentration of Er of 0.01 at .% in these samples even taking into account the experimental error. The latter consideration suggests the presence of low symmetry Er sites in the nanocrystalline ceria samples along with the cubic site though the observed EPR intensities of low symmetry sites are small as compared with cubic sites. Thus, for the low concentration of Er^{3+} centers in studied nanosized and single crystal samples, the coordination number of rare earth ions is mainly 8. Authors concluded that vacancies are located mainly in the next nearest neighbor positions. This finding is not consistent with the conclusion of recent works [12–14] based on DFT calculations that rare earth ion dopants like Er^{3+} trap a vacancy in the nearest neighbor position. It should be noted that at low-level doping it is necessary to consider the impact of uncontrolled impurities from the precursor in cerium

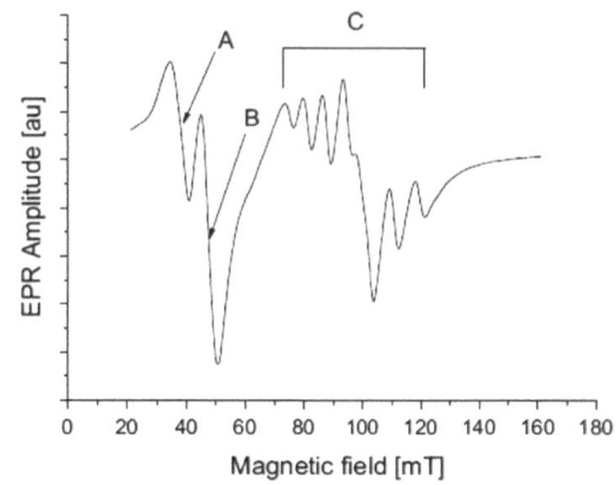

Fig. 9 The EPR spectrum of Sample 1 (crystallite diameter ~22 nm) collected in parallel mode at X-band (9.37 GHz) and 11 K. The lines in the region marked by C are hyperfine signals from Er^{3+} odd isotopes in a cubic environment. The lines marked by A and B are the same as those in Fig. 8. [Reproduced with permission from Appl. Magn. Reson. **46**, 741 (2015)]

dioxide. [Text reproduced with permission from Phys. Status Solidi B **251**, 1545 (2014)]

It was supposed that the weak lines A and B in the spectrum originated from the formation of low-symmetry Er^{2+}; Er^{4+}centers in the lattice. In this case, Er^{2+} should be isoelectronic to Tm^{3+} with the ground state $^3H_6$ and a Landè factor $\Lambda = 7/6$, for which the maximum possible g-value for non-Kramers doublet or two close singlets is $g\| = 2\Lambda 6 = 14$. The EPR line with g ~ 20 (weak line A in Fig. 8) would be to a non-Kramers Er^{4+} ion that is isoelectronic to Ho^{3+}. The ground state of Ho^{3+} is $^5I_8$, with Lande factor $\Lambda = 5/4$ that results in a maximum possible g = 20 for the non-Kramers doublet.

Additional experiments were carried out at X-band with magnetic field B_0 parallel to the microwave magnetic field to demonstrate that the lines at g ~ 14 and ~20 do not originate from Er^{3+} Kramers centers. In this relative geometric configuration of the microwave and the static external magnetic fields the electronic spin transitions are strictly forbidden for Kramers ions. The results for Sample 1 are shown in (Fig. 9). It is clear that the main EPR line from Er^{3+} even isotopes in cubic centers has almost completely disappeared while the hyperfine lines of Er^{3+} odd isotope have become remarkably smaller compared to the lines at A and B, measured also in parallel mode.

3 Room Temperature Ferromagnetism

The recent discovery of room temperature ferromagnetism (RTFM) in oxide nanoparticles such as CeO_2, Al_2O_3, ZnO, In_2O_3, and SnO_2 stimulated studies to find out its origin [15]. They suggested exchange interactions of unpaired electrons trapped in oxygen vacancies (F^+-center) result in RTFM. EPR technique is especially important for studying F-centers in ferromagnetism formation. As applied to ferromagnetism, EPR is most often called electron magnetic resonance, which is successfully used to study magnetic nanoparticles [16–18]. EPR has been used to

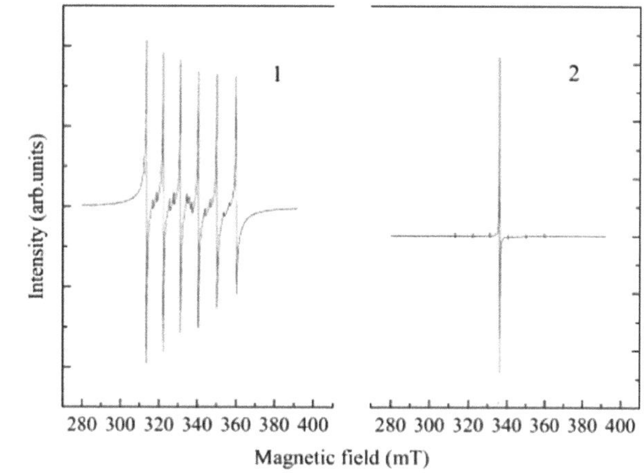

Fig. 10 X-band EPR spectra of the samples at 40 K with grain size of ~35 nm (1) and ~150 nm (2). [Reproduced with permission from Phys. Status Solidi B, (2015). doi 10.1002/pssb.201552542]

study the ferromagnetism in CeO_2 nanocrystals doped with Ni (Ni content more than 1 %) and Co (5 % Co) ion [19, 20]. Askrabic et al. [21] studied pure nanocrystalline CeO_2 synthesized by selfpropagating room temperature method (S-CeO_2 sample) and the precipitation method (P–CeO_2 sample). It was shown that F^+centers with g = 1.99 and g = 1.9 dominated in the S–CeO_2 sample, whereas F^0 centers were the major defects in P–CeO_2 sample. Based on observation of EPR from F^+ centers in the S–CeO_2 sample, that exhibits RTFM, authors could conclude about F^+-center mediated ferromagnetism [22].

Studies of two ceria powder samples with grain sizes of ~35 nm and ~150 nm containing small amount of manganese (0.1 at.%) and traces of iron are reported [23–25]. The ferromagnetism in a CeO_2 powder with a grain size of ~150 nm was found to be substantially weaker than in such a powder with a grain size of about 35 nm. EPR measurements revealed in both samples the lines with g ~2.003 that was assigned to F^+ centers. The comparison of intensities of F^+ center lines (Fig. 10) that was considerably stronger in the sample exhibiting stronger ferromagnetism lead authors to the conclusion that ferromagnetism in studied samples cannot be explained by exchange interacting F^+-centers. The charged transfer ferromagnetism was found to be more suitable to explain experimental results.

4 Toxicological Screening of Ceria Nanoparticles

Cytotoxic effect of Cerium Oxide nanoparticles in several cancer and normal cell lines and their potential to change intracellular redox status is reported [26]. Free radicals related antioxidant capacity of the cells was studied by the reduction of stable free radical TEMPONE radical in the HaCaT, 518 A2, and HT-29 cell lines, untreated and treated with 100 or 200 lM cerium oxide nanoparticles using EPR spectroscopy. The reduction was followed during 20 min since addition of

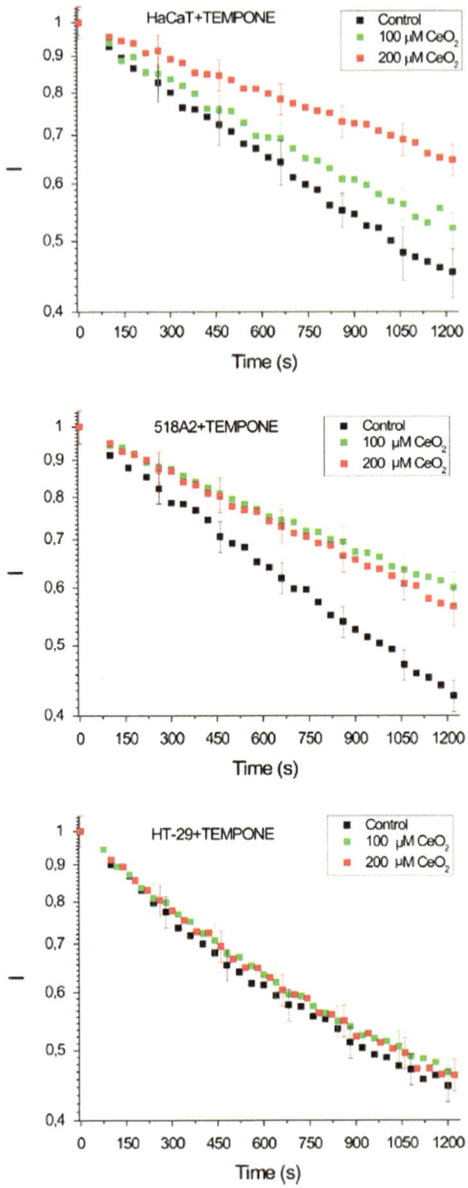

Fig. 11 Reduction of the TEMPONE radical in the HaCaT, 518 A2, and HT-29 cell lines, untreated and treated with 100 or 200 lM cerium oxide nanoparticles. HaCaT cell line has more preserved free radical metabolism. [Reprinted from Chemico-Biological Interactions **232**, 85 (2015), with permission from Elsevier]

TEMPONE to the cells. It could be concluded that HaCaT cell line had more preserved free radical metabolism, and thus redox status, in comparison with the two other cell lines (Fig. 11).

5 Free Radical Scavenging Properties of Ceria Nanoparticles

Radical scavenging properties of ceria nanoparticles may be studied using EPR spectroscopy. For detecting free radicals in chemical and biological system (due to short time), spin-trapping techniques are being used.

EPR of gadolinium doped CeO_2 nanoparticles has been used to assess superoxide radical scavenging using xanthine oxidase/xanthine reaction and spin trap technique [27]. CeO_2 with different dopant concentrations were found to show (Fig. 12) significant scavenging properties. They could find that antioxidant potential decreases with increasing doping. It could therefore be concluded that as doping increases, thus subsequently increasing the Ce(3+)/Ce(4+) ratio, antioxidant potential decreases, suggesting that differences in reactivity of CeO_2 are due to the ability of Ce to transition between the two valence states and the presence of increased oxygen vacancies, rather than dependent on a specific valence state.

Babu et al. [28] used EPR in the presence of a spin trap (5, 5-dimethylpyrroline-N-oxide (DMPO)) to study oxygen free radical scavenging properties of ceria nanoparticle, synthesized by chemical method. The EPR spectra for DMPO-OH adduct and simulated spectra are shown in Fig. 13. Intensity of hydroxyl spectra were plotted with respect to time with and without ceria nanoparticles to find the effect of ceria nanoparticles. The hydroxyl radical

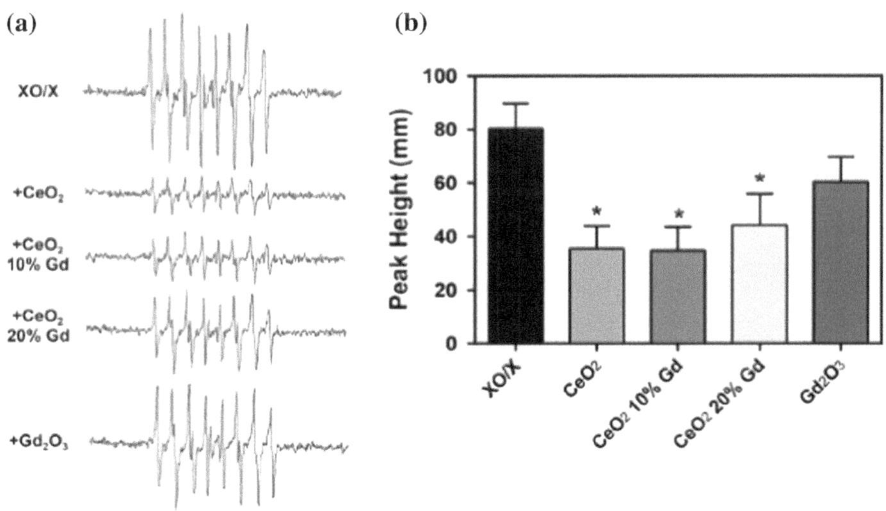

Fig. 12 **a** Superoxide radical scavenging assessed using xanthine oxidase/xanthine reaction and spin trap technique. **b** The first, fourth fifth, and eighth peaks were used for superoxide radical production. All three CeO_2 particles show significant scavenging properties [27]. (Open Access)

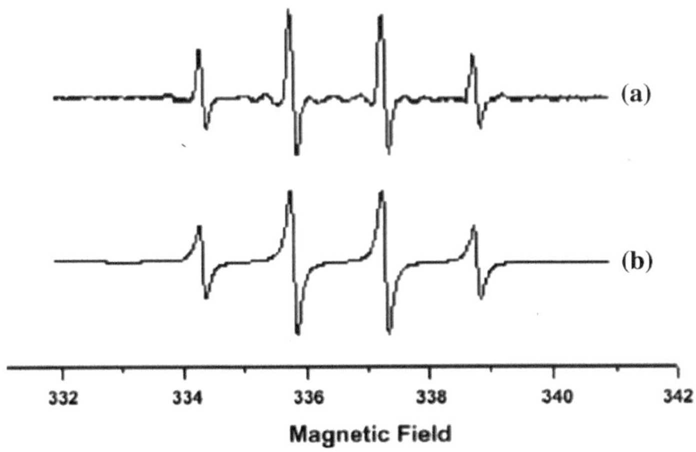

Fig. 13 EPR spectra of DMPO-OH adduct. A represents experimental spectrum and B represents simulated spectrum. Reprinted from Chem. Phys. Lett. **442**, 405 (2007), with permission from Elsevier

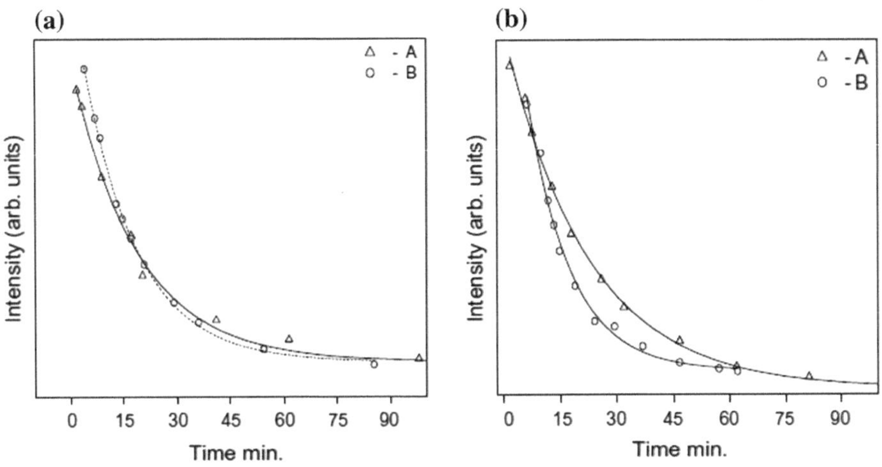

Fig. 14 Extinction plot for hydroxyl radical and hydroxyl radical in presence of 1 mM (**a**) and 10 μM (**b**) concentration of ceria nanoparticles. A corresponds to the intensity of DMPO-OH adduct in the absence and in the presence of nanoparticles, respectively. Intensity of hydroxyl radicals in the presence of ceria nanoparticles decreases very fast at low concentration and hence indicates the possibility to use ceria nanoparticles as free radical scavenger. It also indicates toward the agglomeration at higher concentrations Reprinted from Chem. Phys. Lett. **442**, 405 (2007), with permission from Elsevier

scavenging properties of ceria nanoparticles were found to vary with concentrations (Fig. 14). It was concluded that catalytic effect was better at lower concentration. They presumed that it may be due to the agglomeration at higher concentrations.

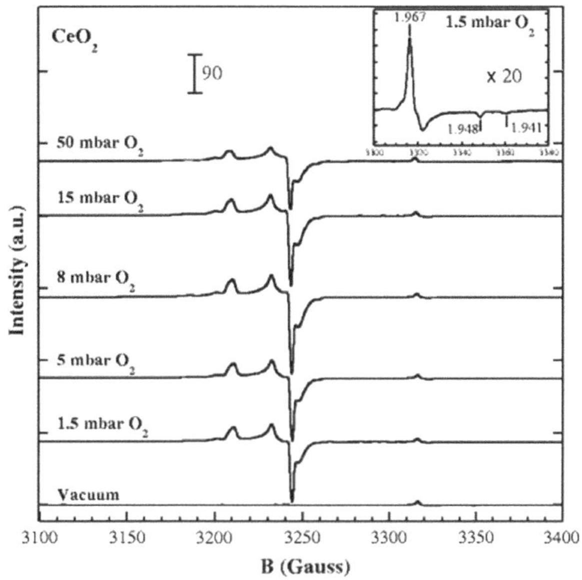

Fig. 15 EPR spectra of CeO_2 sample (77 K) under increasing O_2 pressure [29]. (Open Access)

6 Catalytic Effects of Ceria Nanoparticles

EPR spectroscopy results for oxygen activation property of Gold–ceria (Au/CeO_2) and gold–ceria–alumina ($Au/CeO_2/Al_2O_3$) catalysts are reported by Lakshmanan et al. [29]. Nature and the location of the adsorbed oxygen species could be studied, and they could find that the concentration of superoxo species was lower for oxide-supported gold samples than for bare oxides. Increase in oxygen mobility due to specific gold–ceria interaction is probably responsible for the higher catalytic performance of Au/CeO_2 and $Au/CeO_2/Al_2O_3$ in oxidation reaction compared to bare supports (Fig. 15).

7 Other EPR Applications

Investigation of the extent of ROS generation by different CeO_2 nanoparticles including one prepared by them using hydrothermal method (HT–CeO_2) under different illumination conditions is reported in literature by Lieung et al. [30]. EPR spectra were recorded with DMPO spin trap in 0.9 % NaCl solution and in f2-Si medium on different CeO_2 nanoparticles. In f2-Si medium under UV illumination, significant signals corresponding to hydroxyl radicals were observed. They also further performed ESR measurements to investigate the impurities and suggested that multiple peak signal may be attributed possibly to Gd impurities in CeO_2 [31, 32].

Hernandez-Alonso et al. [33] used EPR spectroscopy to study the photoactivated processes occurring due to UV illumination of CeO_2 nanoparticles. Capability of the samples for the trapping of photogenerated charge carriers has been investigated. The photocatalytic activity of the samples is proposed to be related with the efficiency in the charge separation processes, in accordance with EPR results of oxygen photo adsorption.

EPR assisted by X-ray photoelectron spectroscopy (XPS) and UV-visible analyses is reported which confirms the reactivity of nanoceria as superoxide dismutase (SOD) mimetic [34]. It could be concluded that the surface oxidation state of nanoceria plays an integral role in the SOD mimetic activity of nanoceria. The ability of nanoceria to scavenge superoxide is found to be directly related to cerium (III) concentrations at the surface of the particle.

8 Future Scope

Recent publications about observation of weak room temperature ferromagnetism in pure ceria nanoparticles can give a new impetus to the study of magnetic resonance spectra and explanation of their features. These findings make ceria based material a perspective for application in spintronics. Modifying CeO_2 to enhance its photoactivity in the visible range is another important area which needs further efforts. Toxicological screening also needs attention due to potential applications such as sun creams and outdoor paints leading to increased exposures of ceria nanoparticles.

Acknowledgments We acknowledge publishers of repute for granting us permissions to reuse text extracts and reproduce figures.

References

1. Yahiro, H.Y., Baba, K., Eguchi, H., Arai, J.: J. Electrochem. Soc. **135**, 2077 (1988)
2. Ansari, S.A., Khan, M.M., Ansari, M.O., Kalathil, S., Lee, J., Cho, M.H.: RSC Adv. **4**, 16782 (2014). doi:10.1039/c4ra00861h
3. Marzi, L.D., Monaco, A., Lapuente, J.D., Ramos, D., Borras, M., Gioacchino, M.D., Santucci, S., Poma, A.: Int. J. Mol. Sci. **14**, 3065 (2013). doi:10.3390/ijms14023065
4. Wang, Q., Perez, J.M., Webster, T.J.: Int. J. Nanomed. **8**, 3395 (2013)
5. Araujo, V.D., Avansi, W., de Carvalho, H.B., Moreira, M.L., Longo, E., Ribeiro, C., Bernardi, M.I.B.: Cryst. Eng. Comm. **14**, 1150 (2012)
6. Rakhmatullin, R.M., Aminov, L.K., Kurkin, I.N., Bottcher, R., Poppl, A., Avila-Paredes, H., Kim, S., Sen, S.: J. Chem. Phys. **131**, 124515 (2009)
7. Kim, N., Stebbins, J.F.: Chem. Materials **19**, 5742 (2007)
8. Sameshima, S., Kawaminami, M., Hirata, Y.: J. Ceram. Soc. Japan **110**, 597 (2002)
9. Rakhmatullin, R.M., Aminov, L.K., Kurkin, I.N., Bottcher, R., Poppl, A., Sen, S.: J. Appl. Phys. **114**, 203507 (2013)

10. Rakhmatullin, R.M., Kurkin, I.N., Pavlov, V.V., Semashko, V.V.: Phys. Status Solidi B **251**, 1545 (2014)
11. Rakhmatullin, R.M., Aminov, L.K., Kurkin, I.N., Poppl, A.: Appl. Magn. Reson. **46**, 741 (2015)
12. Andersson, D.A., Simak, S.I., Skorodumova, N.V., Abrikosov, I.A., Johansson, B.: Proc. Natl. Acad. Sci. U.S.A. **103**, 3518 (2006)
13. Nakayama, M., Martin, M.: Phys. Chem. Chem. Phys. **11**, 3241 (2009)
14. Wei, X., Pan, W., Cheng, L., Li, B.: Solid State Ionics **180**, 13 (2009)
15. Sundaresan, A., Bhargavi, R., Rangarajan, N., Siddesh, U., Rao, C.N.R.: Phys. Rev. B **74**, 161306 (2006)
16. Noginov, M.M., Noginova, N., Amponsah, O., Bah, R., Rakhimov, R., Atsarkin, V.A.: J. Magn. Magn. Mater. **320**, 2228 (2008)
17. Noginova, N., Weaver, T., Giannelis, E.P., Bourlinos, A.B., Atsarkin, V.A., Demidov, V.V.: Phys. Rev. B **77**, 014403 (2008)
18. Fittipaldi, M., Sorace, L., Barra, A.L., Sangregorio, C., Sessoli, R., Gatteschi, D.: Phys. Chem. Chem. Phys. **11**, 6555 (2009)
19. Misra, S.K., Andronenko, S.I., Engelhard, M.H., Thurber, A., Reddy, K.M., Punnoose, A.: J. Appl. Phys. **103**, 07D122 (2008)
20. Misra, S.K., Andronenko, S.I., Harris, J.D., Thurber, A., Beausoleil, G.L., Punnoose, A.: J. Nanoscience and Nanotechnology **13**, 6798 (2013)
21. Askrabic, S., Dohcevic-Mitrovic, Z.D., Araujo, V.D., Ionita, G., de Lima Jr, M.M., Cantarero, A.: J. Phys. D Appl. Phys. **46**, 495306 (2013)
22. Paunovic, N., Dohcevic-Mitrovic, Z., Scurtu, R., Askrabic, S., Prekajski, M., Matovic, B., Popovic, Z.V.: Nanoscale **4**, 5469 (2012)
23. Rakhmatullin, R.M., Pavlov, V.V., Semashko, V.V., Korableva, S.L.: ZhETF **148**, 315 (2015)
24. Rakhmatullin, R.M., Pavlov, V.V., Semashko, V.V., Korableva, S.L.: JETP **121**, 274 (2015)
25. Rakhmatullin, R.M., Pavlov, V.V., Semashko, V.V.: Phys. Status Solidi B (2015). doi:10.1002/pssb.201552542
26. Pesic, M., Podolski-Renic, A., Stojkovic, S., Matovic, B., Zmejkoski, D., Kojic, V., Bogdanovic, G., Pavicevic, A., Mojovic, M., Savic, A., Milenkovic, I., Kalauzi, A., Radotic, K.: Chem. Biol. Interact. **232**, 85 (2015)
27. Dunnick, K.M., Pillai, R., Pisane, K.L., Stefaniak, A.B., Sabolsky, E.M., Leonard, S.S.: Biol. Trace Elem. Res. **166**, 96 (2015). doi:10.1007/s12011-015-0297-4
28. Babu, S., Velez, A., Wozniak, K., Szydlowska, J., Seal, S.: Chem. Phys. Lett. **442**, 405 (2007)
29. Lakshmanan, P., Averseng, F., Bion, N., Delannoy, L., Tatibouet, J.M., Louis, C.: Gold Bull. **46**, 233 (2013)
30. Leung, Y.H., Yung, M.M.N., Ng, A.M.C., Ma, A.P.Y., Wong, S.W.Y., Chan, C.M.N., Ng, Y. H., Djurisic, A.B., Guo, M., Wong, M.T., Leung, F.C.C., Chan, W.K., Leung, K.M.Y., Lee, H.K.: J. Photochem. Photobiol. B: Biol. **145**, 48 (2015)
31. De Biasi, R.S., Grillo, M.L.N.: J. Solid State Chem. **178**, 1973 (2005)
32. Figaj, M., Becker, K.D.: Solid State Ionics **141–142**, 507 (2001)
33. Hernandez-Alonso, M.D., Hungria, A.B., Martıinez-Arias, A., Fernandez-Garcıia, M., Coronado, J.M., Conesa, J.C., Soria, J.: Appl. Catal. B **50**, 167 (2004)
34. Heckert, E.G., Karakoti, A.S., Seal, S., Self, W.T.: Biomaterial **29**, 2705 (2008)

Synthesis and Characterization of Undoped and Doped (Mn, Cu, Co) ZnO Nanoparticles: An EPR Study

Şeyda Çolak and Cangül Aktürk

Abstract **Diluted Magnetic Semiconductors** are obtained by introducing dilute amount of transition metal ions into semiconductors and they are subjected to their promising applications to spintronics. **Zinc Oxide (ZnO) nanoparticles,** which are used in large scale both in research and technology, belong to dilute magnetic semiconductors (DMS). In this study, structural, optical, and magnetic characterization of the synthesized undoped and Mn, Cu, and Co-doped (5 %) ZnO nanoparticles by **Co-precipitation method** were performed. Crystal sizes of the synthesized unannealed ZnO nanoparticles have been determined to be **~8 nm** and **~90 nm** by **XRD** and **SEM** analyses, respectively. **AFM** analysis is used for the determination of the surface topographies of the synthesized samples. From the results of **UV-Vis** studies, the absorbance peaks for the synthesized ZnO nanoparticles have been appeared at **~335 nm** and the mean gap energy is calculated to be **3.47 eV. Electron Paramagnetic Resonance (EPR)** spectroscopy is performed for the magnetic characterization of synthesized ZnO nanosamples dominantly and **Vibrating Sample Magnetometry (VSM)** was also used as a complementary technique. No EPR signal has been observed for undoped synthesized ZnO nanoparticles where six resolved EPR resonance lines, four resolved EPR resonance lines, and a broad, single unresolved EPR resonance line was recorded for 5 % doped synthesized ZnO:Mn, ZnO:Cu, and ZnO:Co nanoparticles, respectively. For UV-irradiated and annealed samples, some new EPR resonance lines have also been arised originating from the defects and oxygen damage centers involved in the crystal structure of ZnO. VSM results indicated that none of the synthesized ZnO nanoparticles were purely in ferromagnetic character at room temperature.

Keywords EPR · Nanotechnology · DMS · ZnO · ZnO:Mn · ZnO:Cu · ZnO: Co · VSM · XRD · SEM · AFM · UV-Vis

---

Ş. Çolak (✉) · C. Aktürk
Physics Engineering Department, Hacettepe University, Beytepe, 06800 Ankara, Turkey
e-mail: seyda@hacettepe.edu.tr

A.K. Shukla (ed.), *EMR/ESR/EPR Spectroscopy for Characterization of Nanomaterials*, Advanced Structured Materials 62,
DOI 10.1007/978-81-322-3655-9_7

1 Introduction

1.1 Nanoparticles

Nanoparticles are extremely small in size, which are in the dimension of 100 nm or less. ***Nanotechnology*** is a broad and interdisciplinary research area that has been growing fast worldwide nowadays [3, 24]. Nanotechnology is in relationship with physics, chemistry, biology, computer, and material sciences integrated with engineering.

Nanoparticles have a much greater surface area to their volume ratio which can lead the material to greater chemical reactivity and affect its physical properties when compared with their bulk form [3]. Nanoparticles have created high interest by virtue of some their unusual properties on mechanical, electrical, optical, and magnetic properties and their applications on these areas [3, 26, 45, 53].

Nanotubes, nanowires, and nanobelts are some of the basis of nanoscience and nanotechnology [10] and for future electronic, optical, and optoelectronic nanodevices are aimed to use. Recent developments in nanotechnology and the results of their quantum size effects for nanoscale particles are the indications that most of the devices in the future will be really based on nanomaterials [8, 69]. Besides, magnetic properties of the material can be controlled to an extent just by changing the composition, size, shape, and structure of the nanoparticles [24].

Co-precipitation, sol-gel, thermal decomposition, microemulsion, grinding, hydrothermal, laser ablation, hydrosis, combustion gas-phase reaction, chemical vapor deposition, flame spray, etc, methods are some of the used synthesis mechanisms of nanoparticles [27, 55, 66]. The physical and chemical specifications of the magnetic nanoparticles are strongly depended on the synthesis method preferred [1, 3, 52]. The synthesized ZnO nanoparticles can exhibit the morphologies such as spheres, plates, rod-like, belt, wires, worms-like, tetra-pod, etc.) [13].

1.2 Magnetic Nanoparticles

Magnetic nanoparticles commonly consist of elements such as **iron**, **nickel,** and **cobalt** and their chemical compounds [1, 24]. Magnetic nanoparticles also exhibit some new and interesting physical properties due to their quantum size effects in the nanoscale dimensions, including magnetic interactions, electronic properties, charge transfer, etc. [53].

Ferrite nanoparticles (iron oxide nanoparticles) are in maghemite or magnetite crystal structure and have superparamagnetic character in nanoscale. Superparamagnets exhibit magnetic behavior when an external magnetic field is applied on them. The surface of ferrite nanoparticles are preferred to be modified generally by silica or polymers to increase their stability in solution. Magnetic nanoparticles are used in medical diagnostics and cancer treatments (magnetic hyperthermia), chemistry, biomedical imaging, electronics for storage of information processes, memory applications, and also for a variety of genetics applications [1, 12, 35, 70]. Identification of the magnetic behavior of the widely used nanoparticles such as Zinc oxide is an important concept.

1.3 Dilute Magnetic Semiconductors (DMS)

Semiconductor nanoparticles have physical and chemical properties between molecules and bulk solid semiconductors which are strongly size dependent [21]. Nanostructured semiconductors are known to show quantum size effects such as surface and interface effects, enhanced optical nonlinearity, and excellent photocatalytic properties [8, 26, 69] so that this topic is accepted to be very important and studied across the world [5]. Semiconductors and many other metals can also show some changes in their optical properties depending on their particle sizes. It is known that semiconducting nanoparticles can exhibit fluorescence showing blue shift with the decreasing particle size [11].

Diluted magnetic semiconductors (DMS) obtained by introducing a dilute amount of rare earth or transition metal ions into semiconductors in which the transport and magnetic properties can be combined in a single substance. Thus, DMSs have wide interest recently for their possible applications in generating and manipulating spin-polarized currents [11, 17, 18, 38, 44]. DMSs are subjected to their promising applications particularly to spintronics (spin + electronics) [1, 11, 14, 16, 47, 59]. DMSs have also several favorable properties such as ultraviolet absorbance, piezoelectricity, luminescence at high temperatures, good transparency, high electron mobility, etc. [8].

Recently, investigations on oxide diluted magnetic semiconductors such as transition metal doped ZnO, TiO_2, SnO_2, etc., have been investigated due to their promising magnetic properties [4, 14, 17, 37, 51, 58, 67]. Among the most studied metal oxides nanoparticles, **Zinc Oxide (ZnO)** stands out due to its low cost and ease of synthesis, extraordinary optoelectronical properties, environmentally friendly, and highly versatile device fabrication [57]. Band gap of ZnO nanoparticles demonstrate strong size dependency [19] so ZnO nanoparticles can have broad applications for UV lasers, photodetectors, switches, bio-sensors, solar cells, etc., [62]. ZnO is also a promising host material for ferromagnetic transition metal ion dopants such as Mn, Co, Cu.

1.4 Room Temperature Ferromagnetism of ZnO Nanoparticles

Spintronic technology aims to use the spin of the electron instead of its charge for reading, writing, and transferring data in technology, at high temperatures, especially at room temperature [44]. Room temperature ferromagnetism has been realized from the wide band gap of semiconductors [56]. Ferromagnetism in DMSs is a very interesting problem in magnetism and it is also based on some theoretical work for transition metal doped ZnO nanoparticles [16, 31].

ZnO-based DMSs are accepted to be the possible candidates for high temperature ferromagnetic semiconductors and promising host materials for ferromagnetic doping [1, 6, 19]. Because ZnO-based DMSs have wide band gap (3.4 eV) and large excitonic binding energy (60 meV) to the possibility of achieving room temperature ferromagnetism [51] for the applications such as spin-based light emitting diodes, sensors, and transistors [50].

Above room temperature, ferromagnetism has been predicted for several transition metals doped to ZnO [14] such as Co [4, 9, 61], Mn [19, 67], Fe [19], Cu [6], etc. Although the mechanism responsible for ferromagnetism behavior at high temperature is still not clear [65], it is known that the ferromagnetism of the samples are very sensitive to their synthesization methods or their preparation conditions. So many controversial results, such as very low magnetic property or even the absence of ferromagnetism have been also reported in the literature [1, 11, 14].

1.5 Zinc Oxide (ZnO) Nanoparticles

Zinc oxide is an inorganic compound and it is accepted to be one of the most important semiconductor materials, both with scientific and technological interest [10, 11, 23].

ZnO is present in the Earth's crust as a mineral zincite but mostly it is produced synthetically. ZnO usually appears as a white powder which is nearly insoluble in water and ZnO is used in large scale including plastics, ceramics, glass, paints, pigments, foods (source of Zn nutrient), batteries, ferrites, etc. [8]. ZnO is a II–VI semiconductor where zinc and oxygen belong to second and sixth groups of the periodic table [8]. ZnO crystallizes in three forms such as hexagonal wurtzite, cubic zincblende, and cubic rocksalt but it generally crystallizes in h*exagonal wurtzite* form as this is the most stable form at ambient conditions. The chemical structure of the ZnO is given in Fig. 1.

ZnO has a direct wide band gap energy 3.4 eV besides its high exciton binding energy of 60 meV where the large exciton binding energy provides it more efficient excitonic emission even at high temperatures [19, 54]. It is one of the few oxides that show quantum confinement effects experimentally depending on its size range

Fig. 1 Chemical structure of ZnO

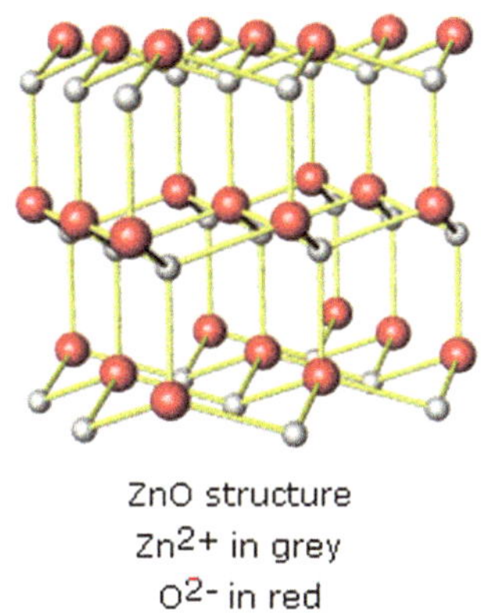

[41]. The quantum confinement of nanocrystalline ZnO particles can be illustrated from the blueshift in their photoluminescence UV emission and the properties of the ZnO nanocrystals depend closely on their particles size, morphology, and surface area [27].

Because of the wide and direct band gap properties, ZnO nanopowders are preferred to be investigated due to their applications in different optic, electronic, spintronic, or magnetic devices [31, 32, 37]. ZnO is also widely preferred in optoelectronic applications such as field effect transistor (FET), transparent electronics and UV light emitters, displays, sensors, ultraviolet and visible lasers, solar cells components, in pharmaceutical and cosmetic industries, in thin film printing, in transparent electronics, in information storage, etc., [5, 6, 8, 13, 18, 20, 22, 25, 39, 40, 42, 43, 45, 47, 49, 52, 54, 55, 59, 62].

1.6 Doped ZnO Nanoparticles

The magnetic properties of wide band gap semiconductors doped with transition metals are another current and controversial topics [6]. Doping is generally used to improve electrical, optical, and mechanical properties of a semiconductor compound [54] and it is needed to use the well-chosen and suitable impurities. ZnO, TiO_2, CeO_2, and SnO_2 diluted magnetic oxides are doped with various transition metals such as Co, Fe, Ni, and Mn and they are investigated both in theoretical and in experimental so as to predict their magnetic properties [1, 6, 11, 20, 31, 44, 66]. It is known that the conductivity and optical properties of ZnO depend strongly on the concentration of doped impurities [45]. As an example, ZnO which is transparent to visible light can be made highly conductive by only doping process [20]. Thus, investigations on doped elements in ZnO host materials and the electrical, optical, magnetic properties of these samples are also very important.

Mn-doped ZnO nanoparticle is very promising due to the wide band gap of ZnO host material and high solubility of Mn atoms in ZnO matrix [14]. Having similar charge state and ionic radius with the host Zn^{2+} cations, Mn^{2+} ions are suitable for sensitive atomic probes in Zn-based compounds [63]. Mn-doped ZnO semiconductors have also become a topic of current interest because of the possibility of their room temperature ferromagnetism, thus Mn-doped ZnO can be a promising material in spintronic applications. Thus, ZnO is one of the host oxide to which addition of Mn causes paramagnetic or ferromagnetic depending on the concentration of the doped Mn and the preparation methods [11, 14, 67]. It has been also observed that doping concentration of Mn has also significant effect on the optical properties of ZnO [31, 39]. For the low doping concentrations of Mn (<3 % mol), the band gap of ZnO is reduced while it is increased for higher Mn doping concentrations (>3 % mol). In addition, a much weaker blue-green peak is also observed at 435 nm and the peak is accepted to be in relation with the surface defects coming through the oxygen vacancies and/or zinc interstitials. As the concentration of Mn increases, the blue emission significantly increased due to the strong exchange interactions in the short range spin system [39].

Cu-doped ZnO nanoparticle is another important research area where metallic Cu and Cu related oxides are nonferromagnetic materials. Some theoretical and experimental studies have confirmed that Cu-doped ZnO nanoparticles can indicate room temperature ferromagnetism depending on their synthesis procedure [6].

Co-doped ZnO nanoparticle has also been intensively studied in aiming to observe the room temperature ferromagnetism in transition metal-doped semiconductors [9]. Especially Co-doped ZnO is possible candidates for high temperature ferromagnetic behavior and has attracted attention [1, 4]. But it is again very sensitive to the synthesization method of the samples and depending on their preparation conditions, thus many controversial results have also been reported.

1.7 Defect Types in ZnO Nanoparticles

The magnetically ordered doped or undoped semiconductors with wide band gaps generally contain a specific amount of point defects. Electronic, magnetic, optical, and structural properties of these semiconductors are very sensitive to the defects where the defects can localize in the metastable energy levels of the wide band gap. The nature of the defects strongly depends on the synthesis method and dimension of the material.

It is evident from literature that various behaviors of ZnO are due to the presence of native defects, such as oxygen vacancies, zinc vacancies, zinc interstitials, and oxygen interstitials [54, 69]. In many cases, these defects act as shallow donors or deep acceptors within the wide band gap of ZnO. Based on ESR and PL analyses, it was concluded that VZn and Zni defects are the main cause of the room temperature ferromagnetism of the ZnO nanostructures [42]. Oxygen vacancies on metal oxide surfaces are found to be electrically and chemically active and responsible for the increasing of the conductivity of the materials [8, 19]. ZnO nanowires are also reported to show n-type semiconductor behavior attributed to intrinsic defects involved in ZnO [19, 22].

These defect centres are also believed to be responsible for visible photoluminescence for ZnO [8, 15] and the concentration of these defects in ZnO nanostructures prepared at low temperatures are believed to be under control by annealing [8]. The presence of OH groups attached to the surface was also accepted to be responsible for the green emission of ZnO [46]. Hydrogen impurities absorbed by ZnO from the air is also reported acting as shallow donors [22]. In another study, core defects are assumed to be originated from the negatively charged zinc vacancies on the crystalline region and shell defects are assumed to be originated from positively charged oxygen vacancies on the surface of ZnO [18]. It is also reported that above 800 °C annealing temperature, ZnO microcrystallites completely become bulk material and no defect centers could be observed [69].

Electron Paramagnetic Resonance (EPR) spectroscopy is a powerful analytical method to monitor these defect centers found and/or created in the samples and can be successfully applied for undoped and doped ZnO nanoparticles [55]. In this study, EPR results for the ZnO nanostructures have also manifested that the Vo centers are the predominant defects in the doped and undoped synthesized ZnO nanoparticles confirming the literature [2].

1.8 Electron Paramagnetic Resonance (EPR) Spectroscopy

Electron Paramagnetic Resonance (EPR) is a very sensitive spectroscopic technique that can detect many types of transition metal ions, free radicals, biradicals, and defect types, particularly for paramagnetic materials [14]. This method was first discovered in 1944, in Kazan University by Zavoisky, on the investigations of electromagnetic energy absorption by paramagnetic metal salts. This technique has high resolution ($\sim 10^{12}$ spins), the measurement is easy, accurate, and harmless to the analyzed sample [68]. EPR crosses multidisciplines including free radicals observed in gaseous, liquid, and solid systems, the irradiated substances in organic compounds, paramagnetic color centers, irradiated chloroplasts, conducting polymers, free radical intermediates, chemical complexes with transition metals, for dosimetric and sterilization purposes, for dating and for spin labeling studies [29].

EPR is also successfully being used for analysis of semiconductor microcrystallites [69]. EPR is mostly playing a key role in the determination of point defects found in bulk semiconductors, for more than 60 years. Thus for the analysis of undoped and doped ZnO nanoparticles, sensitive EPR results are used to be a direct method to monitor the presence of native defects, such as oxygen and zinc vacancies [54, 58, 59].

EPR spectroscopy is based on the absorption of microwave electromagnetic radiation by the unpaired electrons included in the investigated sample, when the sample is placed in an external magnetic field (Fig. 2). When the unpaired electron is placed within the applied magnetic field, H_0, two possible spin states of the

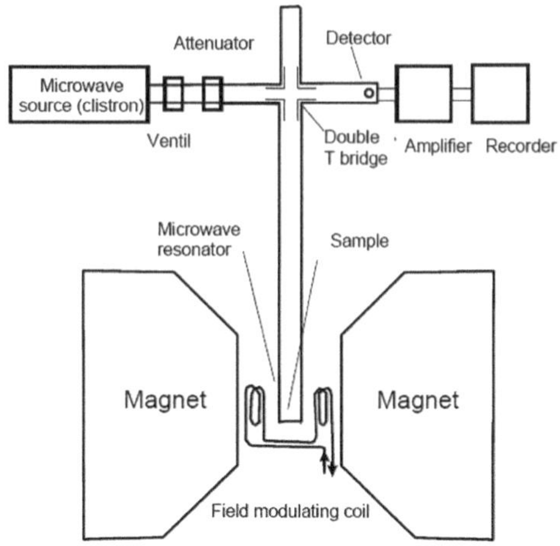

Fig. 2 EPR spectrometer

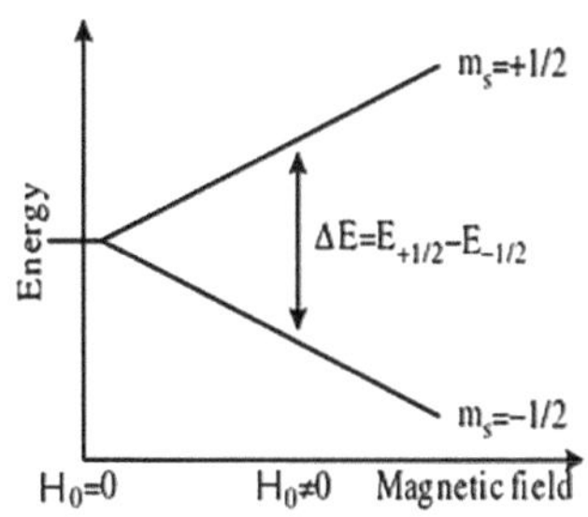

Fig. 3 Energy levels in the existence of external magnetic field

electron will have different energies, by the result of the Zeeman effect. The lower energy state occurs when the magnetic moment of the electron, μ, is aligned with the magnetic field and a higher energy state occurs where μ is aligned against the magnetic field (Fig. 3) [68].

If a second weaker alternating magnetic field H_1 oscillating at a microwave frequency (Fig. 2) is applied at right angles to the static field of H_0, then the electron will begin to precess in the axis of H_1 and when the microwave frequency will be equal to the precession frequency, the ***"resonance condition"*** will be satisfied (Eq. 1). In the Eq. 1; ν is the microwave radiation frequency, h is the Planck constant, g is the spectroscopic splitting factor (g–factor) and β is "Bohr magneton".

$$h\upsilon = g\beta H_0 \tag{1}$$

The unpaired electron absorbs microwave energy in the resonance condition and transitions between the energy states are observed, named as EPR signals. Free radicals have values that are fairly close to $g = 2.0023$ which characterizes a free electron. For crystalline samples, g-factor values corresponding to crystal orientation are designated as g_{xx}, g_{yy}, and g_{zz}, respectively.

EPR signal represents the first-order derivative of absorption line so as to increase the resolution of the recorded spectrum. The area under the line of absorption is proportional to the concentration of paramagnetic centers contained in the sample. The line width of the EPR spectrum gives information about the spin system and the lattice interactions. EPR spectrum can also contains several lines referring to hyperfine splitting relating with the electrons interacting with the neighboring nuclear spins. The magnetic behavior of the electron is modified by its surroundings, so EPR spectra of different samples will be different [68].

EPR provides information not only about the structure, but also the localization of the dopants, particle size, and phases present in the nanostructures with a good accuracy level [63]. It is a very sensitive method for also detecting ferromagnetic ordering as well as the existence of other magnetic species [14, 31, 48, 65].

2 Experimental Results and Discussions

2.1 Synthesis of Undoped and Doped ZnO Nanoparticles

Chemical Co-precipitation Method was used in the synthesis of undoped and *5 % content of Mn, Cu, and Co-doped* samples. Chemical co-precipitation synthesis method is cost effective, has lesser synthesis time, requires low temperature for processing, and both low and homogeneous dimension of nano scale particles can be produced. Chemical synthesis of ZnO nanoparticles has been tried in alcoholic media as ethanol as the growth of oxide particles is slow and controllable in alcoholic media [5].

For the synthesis process of undoped ZnO nano particles, 23 mmol of $Zn(CH_3COO)_2 \cdot 2H_2O$ and 100 mL methanol were stirred by a magnetic stirring for 2 h at 50 °C with 250 rpm. Another solution of 70 mmol of NaOH and 100 ml methanol were stirred for 2 h, at 50 °C. These two solutions were mixed and centrifuged at three steps on the last solution. In the first step, the solution was centrifuged at 8500 rpm for 30 min. For the second step, liquid phase of the solution is separated, ethanol was added and centrifuge was again held at 4500 rpm for 15 min. In the third step, liquid phase of the sample was again separated from the solution, ethanol was again added and it was stirred at 4500 rpm for 5 min. Finally, the liquid phase was separated from the stirred sample and nano sample in solid phase was obtained.

For the synthesis of doped ZnO nano particles; $Mn(CH_3COO)_2 \cdot 4H_2O$, $Cu(CH_3COO)_2 \cdot H_2O$, and $Co(CH_3COO)_2 \cdot 4H_2O$ were added in 5 % content to the prepared solution, for the synthesis of *Mn, Cu, and C* -doped ZnO nano samples, respectively.

2.2 XRD Results

X-ray diffraction is a non-destructive analytical method for identification and quantitative determination of various crystalline forms in powder and solid samples [1, 4, 8, 27, 45, 66]. For the investigation of structural properties of undoped and 5 % Mn^{2+}, Cu^{2+}, and Co^{2+} doped ZnO samples, XRD studies were held.

Rigaku D Max-2200 (CuKα-40 kV, 40 mA, $200V_{AC}$, 3 phase) type of XRD diffractometer was used in XRD experiments. Crystallite size of the samples were found by using Scherer's formula (Eq. 2) where in the equation t is the crystallite size, λ is wavelength ($\lambda = 1.5405$ Å), β is the full maxima half width, θ is diffraction angle and k is a constant ($k = 0.9$).

$$\beta \cos \theta = \frac{\lambda k}{t} \tag{2}$$

XRD results indicated that Mn, Cu and Co ions in the ZnO nanoparticles only substituted with Zn ions (Figs. 4 and 5). The synthesized undoped and doped ZnO nano samples have the wurtzite structure and no secondary phases were detected although some little amount of impurities were observed and these results are also agree with the literature [1, 4, 27, 34, 38, 44, 45, 56, 64, 66].

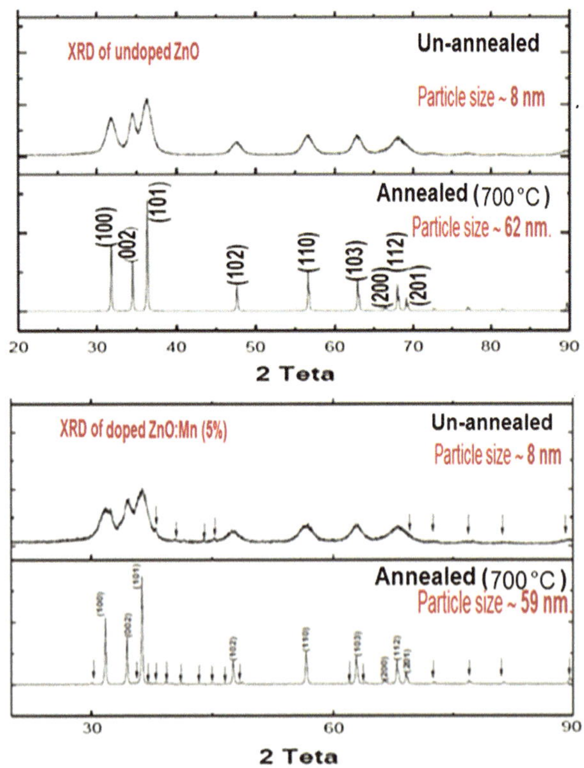

Fig. 4 XRD results of synthesized ZnO nanoparticles (unannealed and 700 °C annealed undoped and Zn:Mn doped (%5) samples)

Fig. 5 XRD results of synthesized ZnO nanoparticles (unannealed and 700 °C annealed Zn:Cu and ZnO:Co-doped (%5) samples)

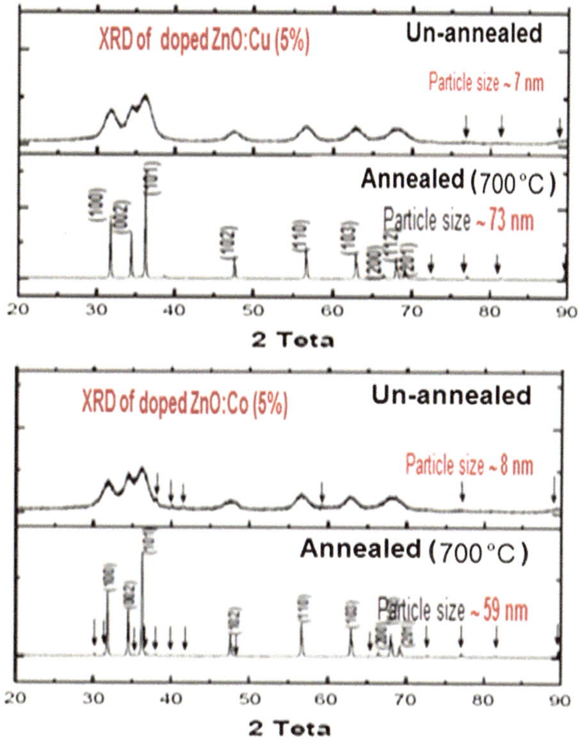

Annealing studies for the synthesized samples were held at 300 °C (1 h), 500 °C (2 h) and 700 °C (2 h). For the annealing studies, MF 120 ash furnace with a Ni–CrNi temperature controller [600–1200 °C (±2 °C)] has been used. The samples were wet after the annealing procedure and white color of the samples were turned to pink, gray, or black colors after annealing process. Thus, organoleptic properties of the samples were changed after the annealing procedure which is also confirmed by some of the data in the literature [7, 69].

Annealing procedure causes to decrease the defects found in the samples, thus optical, electrical, and structural properties of the samples are expected to be cured at high temperatures [17]. In this study; the particle dimensions of undoped and doped ZnO nano samples have observed to be increased by the treatment of heat and the broad peaks recorded for the unannealed samples became sharper for the annealed samples (Figs. 4 and 5). The average of the particle size of the unannealed samples were found to be ~8 nm by using Scherer formula (Eq. 2) where the average particle size was ~60 nm for the samples annealed at 700 °C. These dimensions are also in good agreement with SEM results.

2.3 *Scanning Electron Microscope (SEM) Results*

Scanning Electron Microscope (SEM) analyses were recorded by using an Carl-Zeiss EVO-50 EP Scanning Microscope (W flaman-20 kV–10 pA, gain is 65,000–150,000). The dark shapes in the SEM results generally related with the rich content of Zn and doped ions (oxygen content is low). It is observed that the unannealed and annealed doped ZnO nano samples were nearly cylindrical in shape with particle sizes of **~90** and **~200 nm**, respectively (Fig. 6). Agglomerations

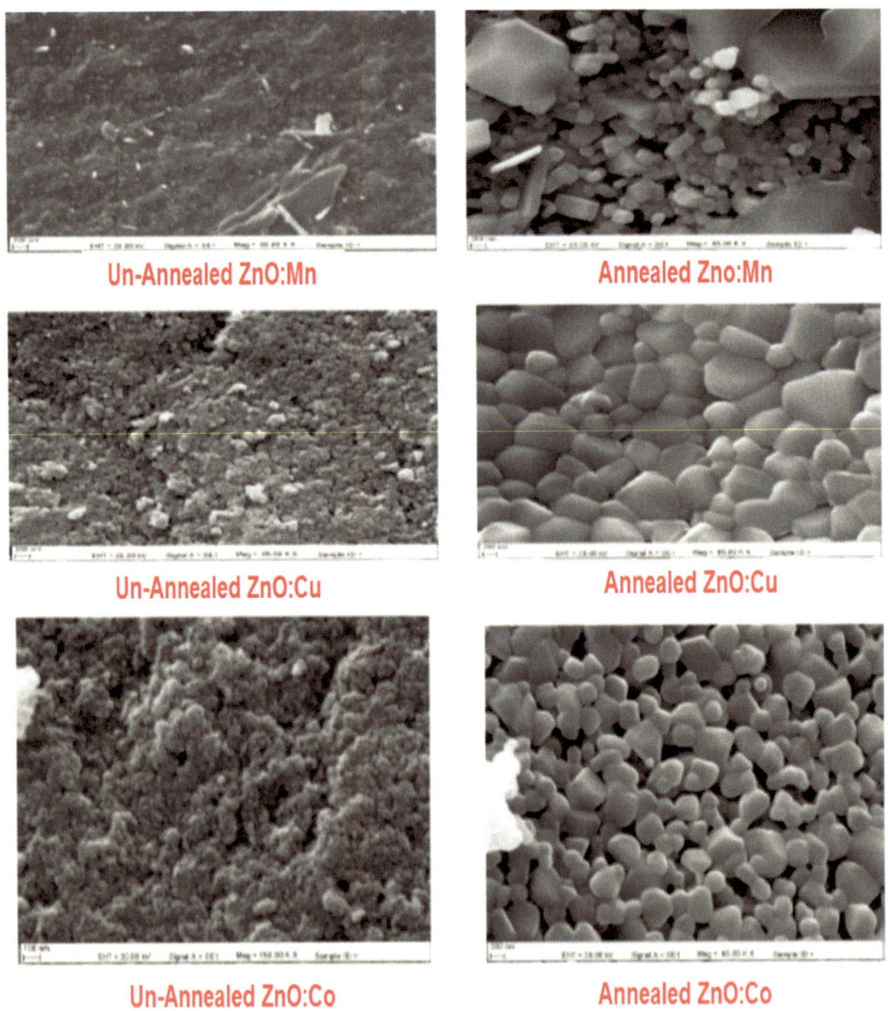

Fig. 6 SEM images of synthesized undoped and doped ZnO nanoparticles

have been also taken place for the 700 °C annealed doped ZnO nanoparticles which is in agreement with the XRD results obtained. This result is also confirmed with literature [10, 67].

2.4 Ultraviolet-Visible (UV-Vis) Spectroscopy Results

In Ultraviolet-Visible (UV-Vis) spectroscopy, high energy electromagnetic radiation in the wavelength range of 100–700 nm is utilized to promote electrons to higher energy orbitals. Since orbitals have quantized energy, only certain transitions can occur in the UV-Vis energy range. The differences in the incident and transmitted beam give us information about the frequencies which are absorbed by the investigated samples. Basing on the absorbance data recorded, the chemical structure of the investigated sample can be analyzed [12]. Among various spectroscopic methods, UV-Vis spectroscopy has been often used to characterize the optical properties of the nanoparticles.

ZnO is accepted to be excellent promise for blue and ultraviolet optical devices because of its wide band gap energy [20, 27] and the quantum confinement of nanocrystalline ZnO particles is illustrated from the blueshift in its photoluminescence UV emission [27]. The band gap energy is also changed by doping type and content to ZnO nanoparticles [14, 20, 39]. In this study, UV-Vis is used to get information about the absorption curve of doped and undoped synthesized ZnO nanoparticles, besides the calculation of the band gap energy of the samples.

For identification of the optical behavior of the synthesized undoped and doped ZnO nanoparticles, a Hitachi U-1900 model of UV-Vis spectrometer has been used. The synthesized samples that were handled before centrifuge process (in liquid form) and stored at normal conditions (atmospheric pressure and moisture) in dark were used for UV-Vis studies.

The strong UV absorbance peaks were appeared approximately at $\boldsymbol{\lambda_{average}}$ **= 335 nm** and the mean gap energy was calculated to be $\mathbf{E_{gap}}$ **= 3.47 eV**

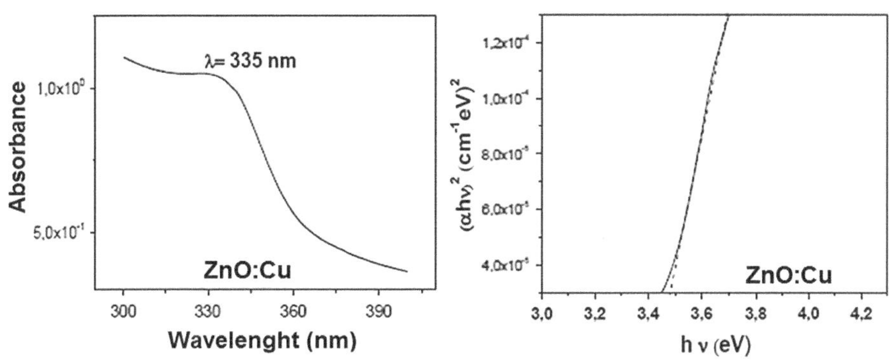

Fig. 7 UV-Vis result of synthesized ZnO:Cu nanoparticle

by using ***Beer-Lambert's law.*** These results are in consistent with also the literature [5, 13, 40, 54]. UV-Vis result of ZnO:Cu sample is given in Fig. 7. It is observed that with the decreasing of particles sizes, UV light absorbance of the ZnO nanoparticles was increased. Thus, the optical properties get better with the increasing of surface-to-volume ratio for Zinc oxide nanoparticles [43].

2.5 Atomic Force Microscope (AFM) Results

The surface topographies of the synthesized Mn-doped ZnO nano particles are investigated before and after annealing process by AFM studies. Nanomagnetics Instruments Ambiant (1 μm × 1 μm area, 1 μm/s sweep velocity, 256 × 256 pixel resolution, dynamic mode) AFM instrument was used in the experiments. Different growing directions and the grain sizes were observed from AFM results for unannealed and annealed ZnO:Mn nanoparticles (Fig. 8). Agglomerations have also been observed from the AFM images of the samples which are in agreement with the literature [34, 67].

2.6 Electron Paramagnetic Resonance (EPR) Results

Electron Paramagnetic Resonance (EPR) spectroscopy is a quick, nondestructive analytical method for investigation of particulary the paramagnetic centers in the samples. In this study, the magnetic characterization of synthesized ZnO nano samples has been performed dominantly by using electron paramagnetic resonance (EPR) spectroscopy. Vibrating sample magnetometry (VSM) was also used as a complementary technique.

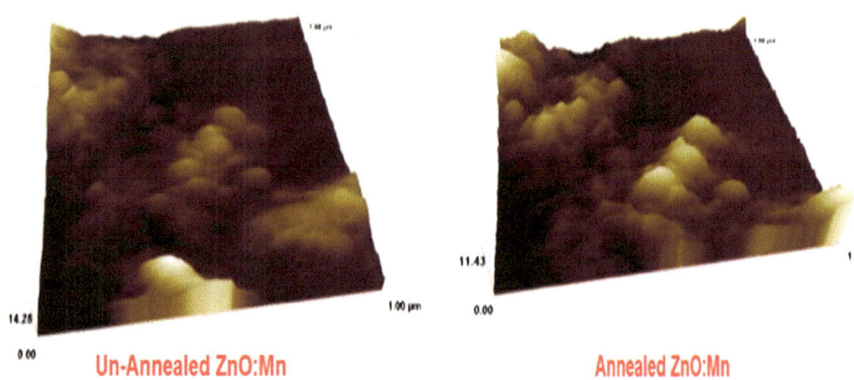

Fig. 8 AFM results of unannealed and 700 °C annealed ZnO:Mn nanoparticles

Bruker EMX-131 X-band (9.3 GHz) EPR spectrometer was used in the experiments. ***Spectrometer operating conditions were as follows***; central field: 348.5 mT; sweep field: 200 mT; microwave frequency: 9.84 GHz; microwave power: 1 mW; modulation frequency: 100 kHz; modulation amplitude: 1 G; receiver gain: 6.3×10^4; sweep time: 81.92 ms; time constant: 40.9 ms.

2.6.1 EPR Results for Undoped Synthesized ZnO Nanoparticles

The ***defect types*** found in nano scale ZnO is generally related with the oxygen vacancies Vo^{+2} (O^{-2} superoxide ions, F centers) in the literature and this defect type is accepted to be responsible from the EPR signal for the undoped ZnO nanoparticles [7, 18, 28]. Vo^{+2} oxygen vacancy is not stable and appears from the shallow part of the ZnO nanoparticles and it gives rise to the EPR resonance line with the corresponding g-value of 2.0021 [30] which is very close to the g-value of the free electron ($g_{\text{free electron}} = 2.0023$). Vo^{+2} oxygen vacancy can also be trapped by a single electron (Vo^{+1}) or by two electrons (Vo^0). Vo^{+1} oxygen vacancy (O^{-1} ion) is contributing from the deeper part of ZnO (core of ZnO) nanoparticle which is more stabile and giving rise to the EPR resonance line with the corresponding g-value of 1.9588 [28]. Zn vacancies found in the structure of ZnO nanoparticles are also accepted to be relatively stable defect type.

In this study, no EPR resonance signal was detected for the synthesized undoped ZnO nanoparticles, indicating that undoped ZnO nanoparticle was in diamagnetic nature. When the undoped ZnO nanoparticles were annealed at 300 °C for 1 h, a single, broad, and unresolved EPR resonance line with spectral parameters of $g = 2.169$ and $\Delta H_{pp} = 6$ mT was arised. Besides, two other resonance lines (I_a and I_b), hardly distinguishable from noise, with corresponding g-values of $g_a = 2.009$ and $g_b = 1.959$ were also observed. I_a and I_b resonance lines are accepted to be originating from Vo^{+2} and Vo^{+1} damage centers, respectively (Fig. 9). For the samples annealed at 500 °C for 2 h, the intensities of I_a and I_b resonance lines increased while the area calculated for the whole EPR spectrum recorded had been decreased. For the annealed samples at 700 °C for 2 h, a dramatic decrease for the whole spectrum, including I_a and I_b resonance lines, had been observed, indicating the sharp decrease in the number of paramagnetic centers found in the sample. But one of the new resonance line I_b ($g_b = 1.959$) could be still distinguishable from the noise. This result indicates that I_b resonance line was related with Vo^{+1} damage center which has deeper (core) location in ZnO crystal structure and is relatively stable [28].

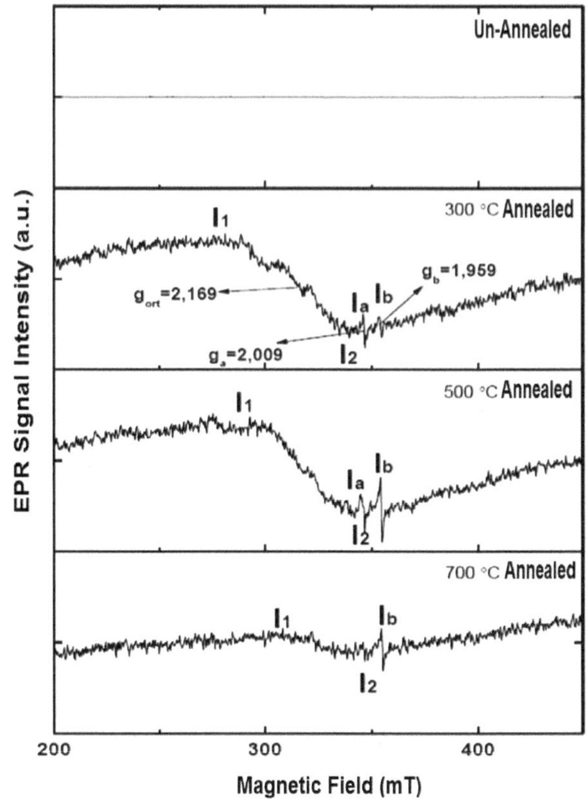

Fig. 9 EPR spectra of unannealed and annealed (300, 500, and 700 K) synthesized undoped ZnO nanoparticles

2.6.2 EPR Results for Mn-Doped Synthesized ZnO Nanoparticles

Mn^{+2} ion has $3d^5$ electronic configuration, electronic ground state of $^6S_{5/2}$, and spectroscopic splitting factor of $g \sim 2.002$ [68]. The variation of Mn concentration (1, 2 and 5 %) have not changed the EPR spectrum pattern of the Mn-doped ZnO nanoparticles but the EPR signal intensities have increased with increasing the Mn content. For the 5 % Mn-doped ZnO nanoparticles, 6 resolved EPR resonance lines were recorded and *g*-value related with the central EPR resonance line was calculated to be 2.009. The recorded spectrum was spread over a 80 mT magnetic field range and the average hyperfine splitting factor for ZnO:Mn nanoparticles was calculated to be 8.0 mT (Fig. 10).

The baseline signal observed in the spectrum which is shown with dashed line is related with the "*ferromagnetic resonance behaviour*" of ZnO:Mn nanoparticles, where six resolved narrow ESR signals are related with the "*paramagnetic character*" of Mn^{+2} ions found in ZnO:Mn nanoparticles. Thus, for Mn-doped ZnO nanoparticles, both *paramagnetic and ferromagnetic characters* are observed

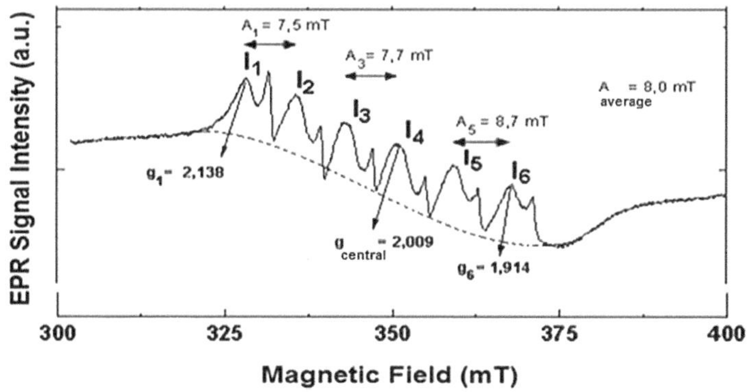

Fig. 10 EPR spectrum of 5 % Mn-doped ZnO nanoparticle

together which is also confirmed by VSM results in this study. There are also similar results appeared in the literature [14, 16, 36].

In annealing studies, no EPR spectrum pattern change had been observed for Mn-doped ZnO nanoparticles but the average value of hyperfine splitting constants

Fig. 11 EPR spectra of unannealed and annealed (300, 500, and 700 K) synthesized %5 Mn-doped ZnO nanoparticles

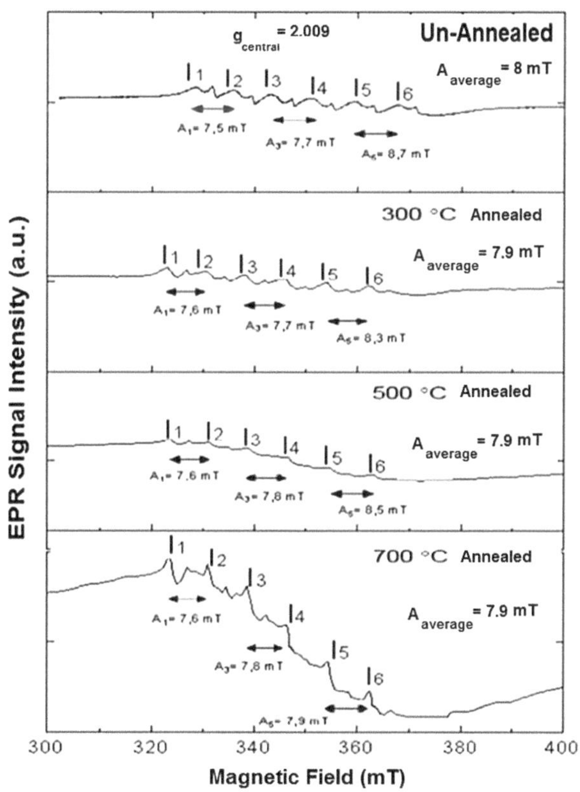

calculated for the unannealed samples (8.0 mT) were decreased in small ratio (7.9 mT) when the samples were annealed. The number of paramagnetic centers contributing to Mn-doped ZnO nanoparticles annealed at **300 °C for 1 h** was calculated by double integration method and it is seen that the area of the spectrum which is linearly proportional with the number of magnetic centers involved in the sample had increased in the order of ~ 10 %. Magnetic centers contributing to the recorded EPR spectrum continued to increase for also the samples that were annealed at **500 °C for 2 h** and finally the number of magnetic centers calculated for the annealed samples at 700 °C were calculated to be ~ 10 times higher than the case for unannealed samples showing that the magnetic unit content was increased for the heat treated ZnO:Mn nanoparticles (Fig. 11). This result was also confirmed by VSM data.

2.6.3 EPR Results for Cu-Doped Synthesized ZnO Nanoparticles

Cu^{+2} ion has $4s^1$ electronic configuration, electronic ground state of $^2D_{5/2}$, and spectroscopic splitting factor with g ~ 2.116 [68]. For the 5 % Cu-doped ZnO nanoparticles, the g-value corresponding to the central resonance line of EPR spectra was found to be 2.063 where four resolved EPR resonance lines were observed. The hyperfine splitting factor was calculated to be A = 13.5 mT and the spectrum was spread over 100 mT magnetic field range (Fig. 12).

Annealing process caused sharp decreases in the intensities of EPR resonance lines recorded for the 5 % Cu-doped ZnO nanoparticles. This result indicates that the number of magnetic centers found Cu-doped ZnO nanoparticles decreases under

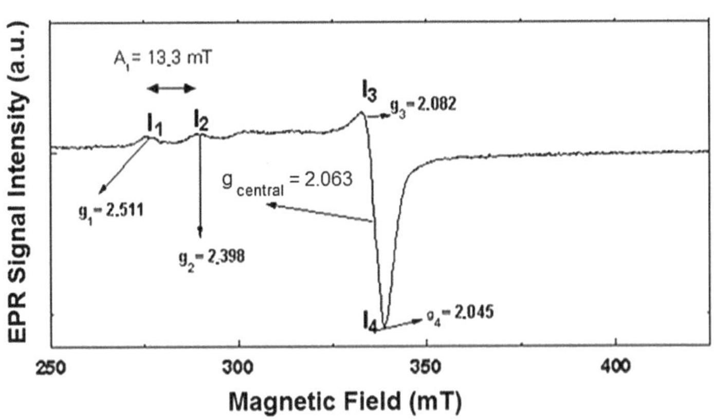

Fig. 12 EPR spectrum for 5 % Cu-doped ZnO nanoparticle

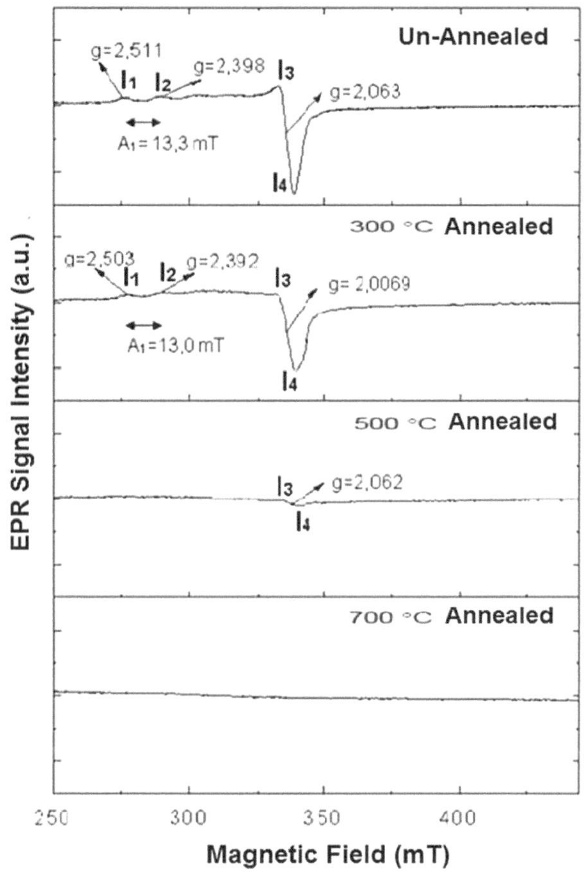

Fig. 13 EPR spectra of unannealed and annealed (300, 500, and 700 K) synthesized Cu-doped ZnO nanoparticles

the influence of heat treatment (Fig. 13) and ZnO:Cu nanoparticles annealed at 700 °C were completely in *diamagnetic* character. This result was also confirmed by VSM studies.

2.6.4 EPR Results for Co-Doped Synthesized ZnO Nanoparticles

Co^{+2} ion has $3d^7$ electronic configuration, electronic ground state of $^4F_{9/2}$, and spectroscopic splitting factor with $g_{//} \sim 2.2$ and $g_{\perp} \sim 4.6002$ [51, 68]. A single, unresolved, broad EPR resonance line was recorded for 5 % Co-doped ZnO nanoparticle, and the spectral parameters such as *g*-value and peak-to-peak width were calculated to be 2.119 and 77.3 mT, respectively (Fig. 14).

A smooth increase in the EPR signal intensity has been observed for Co-doped ZnO nanoparticles which are annealed at 300 °C for 1 h. Besides, one more EPR

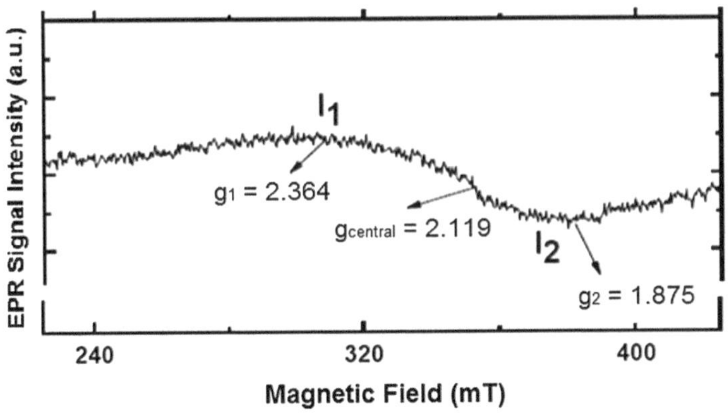

Fig. 14 EPR spectrum of 5 % Co-doped ZnO nanoparticle

resonance line (I_a) with g_a value of 2.002, that is related with Vo^{+2} center, had also been observed (Fig. 15). The whole spectrum area (double integration method is used) recorded for the annealed sample at 500 °C for 2 h was found to increase in the order of ~%3 where the intensity of new appeared I_a resonance line was

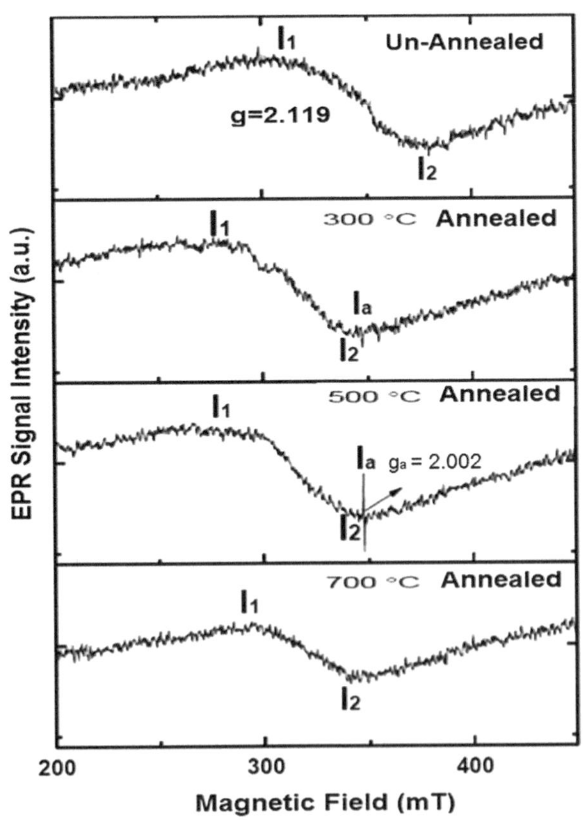

Fig. 15 EPR spectra of unannealed and annealed (300, 500, and 700 K) synthesized 5 % Co-doped ZnO nanoparticles

increased ~10 times compared with that of annealed at 300 °C. But a sharp decrease has been observed for the annealed samples at 700 °C for 2 h and besides I_a resonance peak had disappeared completely.

2.6.5 EPR Results of UV-Irradiated Undoped and Doped ZnO Nanoparticles

To examine the effect of UV-irradiation on the synthesized undoped and doped ZnO nanoparticles, the samples were exposed upon UV-irradiation with *additive dose method* for predetermined irradiation times up to 200 min. The samples were UV irradiated with a distance of 60 cm to irradiation source with using a Termo Oriel-6295 model, 1000 W monochromatic Hg (Xe) arc lamp.

UV irradiation caused some organoleptic changes during the irradiation process, such as darkening colors of the irradiated samples. This result is in agreement with literature where such similar effects have also been obtained by Schneider et al. [59].

No EPR signal was observed for UV irradiated undoped ZnO nano particles but two new EPR resonance lines I_a and I_b with corresponding g-values of $g_a = 2.009$ and $g_b = 1.959$ have been arised in the EPR spectrum of UV-irradiated undoped ZnO nanoparticles. These new resonance lines were probably causing from the oxygen damage centers involved in the crystal structure of ZnO nanoparticles.

For the UV irradiated Cu-doped ZnO nanoparticles, a new EPR resonance line, I_a, with $g_a = 2.003$ had also been appeared for the UV-irradiated samples, at the irradiation duration of 30 min. The intensity of this new resonance line was increased linearly with increasing the irradiation time, up to 200 min. This I_a resonance line is accepted to be related with Vo^{+2} damage center and this result is also in agreement with the literature [54].

For UV-irradiated Co-doped ZnO nanoparticles, another new EPR resonance line, I_a, with $g_a = 2.001$ had appeared after 5 min of UV-irradiation duration and the line intensity of this new resonance line, which is related with Vo^{+2} damage center, had increased slightly with the increasing UV-irradiation time. The variations of I_a resonance line with UV-irradiation time are not given.

Oxygen and zinc vacancies can exist in more than one charge state, and they are often using to explain the optical properties of ZnO. Exposure of light can produce new EPR resonance lines on the irradiated ZnO nanoparticles originating from oxygen vacancies and/or impurities related with zinc vacancies [33]. It is observed that irradiation at room temperature with high energy electrons (3 meV) can also produce oxygen and zinc vacancies in the crystal structure of ZnO nanoparticles [33].

The EPR spectrum area variations of the UV-irradiated Mn, Cu, and Co-doped ZnO nanoparticles with UV-irradiation time is given in Fig. 16. It is seen from this

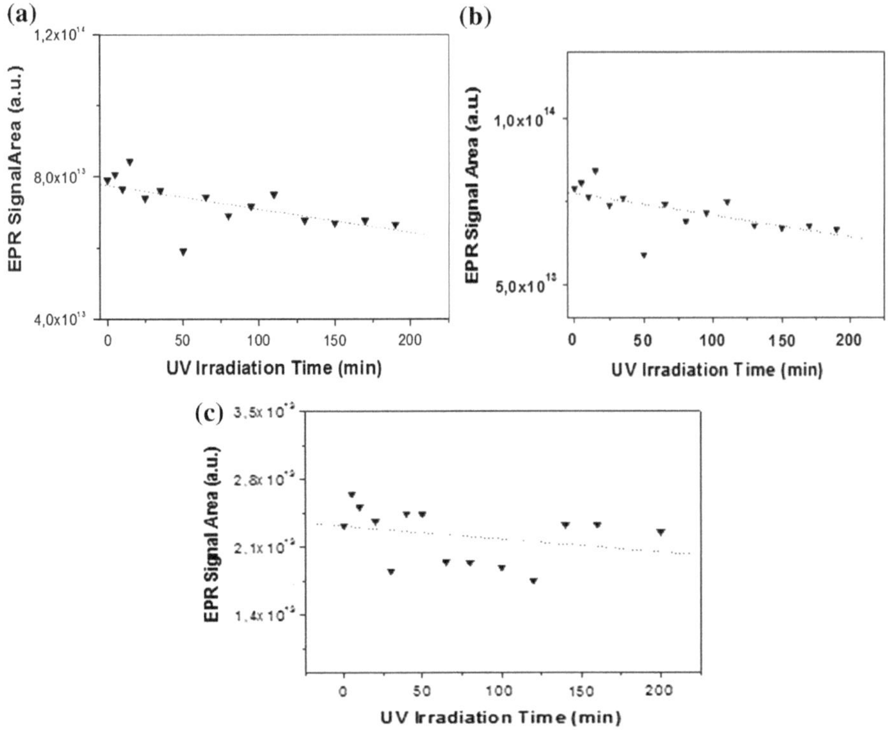

Fig. 16 EPR signal intensities of UV-irradiated 5 % doped ZnO nanoparticles for different irradiation times. **a** Mn-doped ZnO, **b** Cu-doped ZnO, **c** Co-doped ZnO

figure that the area, relating with the magnetic units found in doped ZnO nanoparticles, had decreased slightly with the increasing UV-irradiation time (Fig. 16).

2.7 VSM Results

It is known that obtaining room temperature ferromagnetism will greatly advance future research [49, 66]. ***Vibrating sample magnetometry (VSM)*** technique was used as an alternating and complementary method for the magnetic characterization of synthesized undoped and doped ZnO nanoparticles. For VSM analysis, Cryogenic Limited PPMS, and ADE Magnetics VSM measurement systems were used.

It was observed that the undoped and unannealed ZnO nanoparticles were in diamagnetic character (Fig. 17). The VSM results obtained for undoped and annealed at 700 °C (for 2 h) ZnO nanoparticles showed also **diamagnetic** behavior

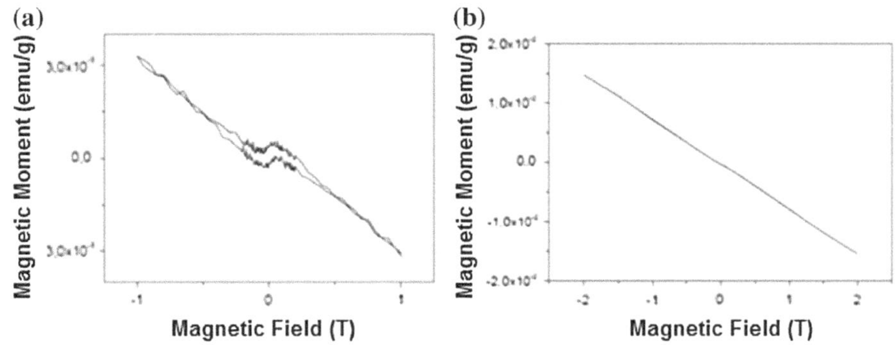

Fig. 17 VSM results of undoped ZnO. **a** unannealed, **b** annealed at 700 °C

confirming the results of EPR studies. Similar results were also reported in the literature depending on the preparation conditions of ZnO nanoparticles [30, 56].

VSM result for %5 Mn-doped unannealed ZnO nanoparticles showed that this sample indicated both **ferromagnetic** and **paramagnetic** character together (Fig. 18). But the annealed Mn-doped ZnO nanoparticles at 700 °C showed only **paramagnetic** type of magnetic character. The same results were also observed by EPR spectroscopy.

In the literature, some controversy types of magnetic characters for Mn-doped ZnO nanoparticles have been observed [67]. Sharma et al. [60] have observed ferromagnetism above room temperature for bulk and thin film forms of Mn-doped ZnO. In another study, the ferromagnetism in Mn-doped ZnO nanoparticles have appeared for low temperature annealed samples and disappeared in the samples annealed at high temperatures [65]. But there are also some several reports on Mn-doped ZnO samples that do not exhibit ferromagnetic properties. The ferromagnetism of such materials is accepted depending strongly on their synthesis

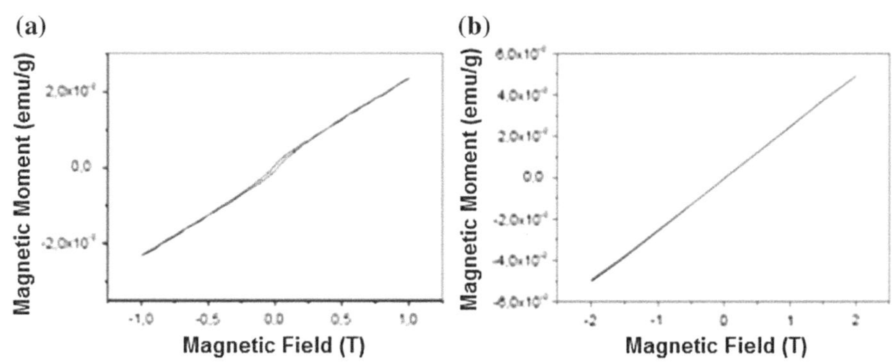

Fig. 18 VSM results of Mn-doped ZnO nanoparticles. **a** unannealed, **b** annealed at 700 °C

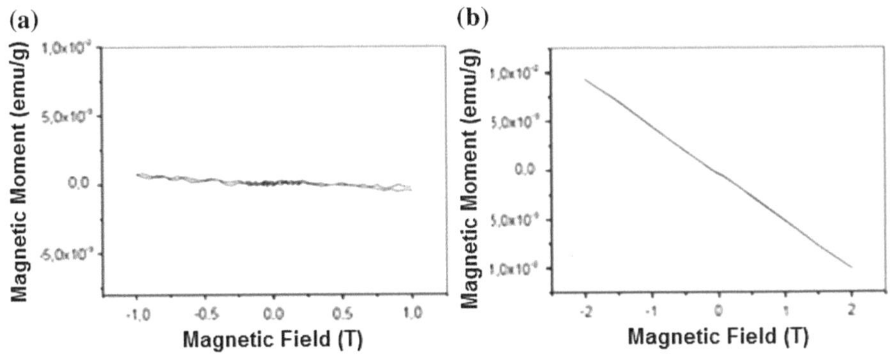

Fig. 19 VSM results of Cu doped ZnO nanoparticle. **a** unannealed, **b** annealed at 700 °C

procedures preferred [48]. The difference in the ferromagnetic properties of the Mn-doped ZnO nanoparticles can be also attributed to the Mn content [11].

VSM result recorded for 5 % Cu-doped ZnO nanoparticles which was unannealed and annealed at 700 °C indicated **diamagnetic** characters. It is well known that this property is strongly dependent on the preparation conditions of the samples and the type of synthesis method used (Fig. 19).

VSM result of 5 % Co-doped unannealed ZnO nanoparticles showed that this sample had both **ferromagnetic** and **paramagnetic** type of magnetic property. The annealed Co-doped ZnO nanoparticles at 700 °C were found to be in **paramagnetic** character by using VSM studies (Fig. 20). These results were also confirmed by the EPR data obtained for the same sample. Ferromagnetic character of ZnO:Co nanoparticle at room temperature is observed in the literature [61].

Room temperature ferromagnetism in doped samples is also accepted to be generated from the oxygen vacancies in the structure of ZnO [14]. There are some studies in the literature indicating that doped ZnO nanoparticles with Co, Mn, Ni, etc., can be ferromagnetic at room temperature [1, 17] but this character is still in

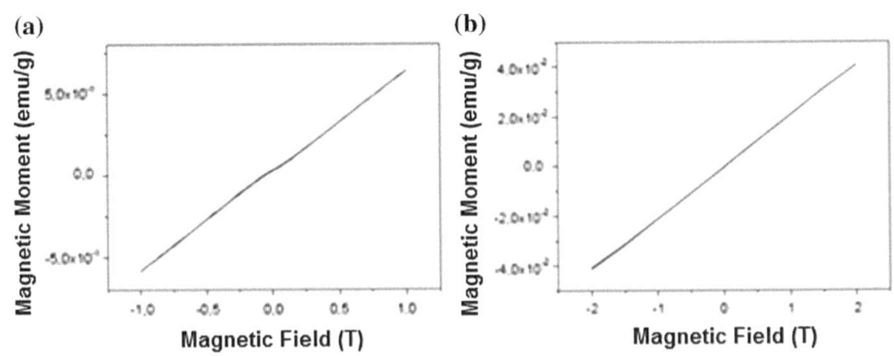

Fig. 20 VSM results of Co-doped ZnO nanoparticles. **a** unannealed, **b** annealed at 700 °C

debate [9] as the magnetic properties of doped ZnO are most likely dependent on the preparation conditions [11, 14]. There are also some studies in the literature where ZnO nanoparticles indicating bulk character after the annealing process [7, 30] as annealing process decreases the impurities of defect centers.

3 Conclusion

In this study; structural, optical, and magnetic properties of undoped and 5 % content of Mn, Cu, and Co-doped ZnO nanoparticles were synthesized by chemical co-precipitation method and have been investigated dominantly by EPR, and some other complementary techniques such as XRD, SEM, AFM, UV-Vis, and VSM. Crystal sizes of the synthesized ZnO nanoparticles have been determined to be ~8 nm by XRD analysis and particle sizes of the ZnO nanoparticles were found to be ~90 nm by SEM analysis. Annealing process increased the dimensions of the synthesized nanoparticles. The surface topographies of the samples were recorded by AFM. UV-Vis results indicated that for undoped and doped synthesized ZnO nanoparticles, the absorbance peaks were appeared at about $\lambda = 335$ nm and the mean gap energy is calculated to be ~3.47 eV. No EPR signal had been observed for undoped ZnO nanoparticles. For ZnO:Mn; six resolved EPR resonance lines have been recorded with $g_{central} = 2.0097$, for ZnO:Cu; four resolved EPR resonance lines have been recorded with $g_{central} = 2.0637$ and for ZnO:Co; a single broad resolved EPR signal have been recorded with $g = 2.1199$. UV irradiation and annealing processes caused some organoleptic changes on the samples and some new EPR resonance lines were also appeared in their EPR spectra with very low intensities and corresponding g-values of $g_a = 2.0021$ and $g_b = 1.9588$, probably causing from the oxygen vacancies (damage centers) involved in the crystal structure of ZnO. VSM findings have indicated that, unannealed and annealed undoped ZnO were diamagnetic, unannealed, and annealed Mn-doped ZnO nanoparticles were paramagnetic + ferromagnetic and paramagnetic, unannealed and annealed Cu-doped ZnO nanoparticles were diamagnetic and unannealed and annealed Co-doped ZnO nanoparticles were in paramagnetic + ferromagnetic and paramagnetic.

References

1. Ahmed, F., Kumar, S., Arshi, N., Anwar, M.S., Koo, B.H., Lee, C.G.: Doping effects of Co^{2+} ions on structural and magnetic properties of ZnO nanoparticles. Microelectron. Eng. **89**, 129–132 (2012)
2. Aktürk, C.: Synthesize of ZnO Nano Systems and Investigation of their Magnetic Properties by Using Electron Paramagnetic Resonance (EPR) Spectroscopy, 2013, M.Sc. Thesis, Hacettepe University, Ankara, Turkey

3. Alagarasi, A.: Introduction to Nanomaterials, Chennai, 2011, India: National Centre for Catalysis Research, 1–76
4. Archer, T., Pemmaraju, C.D., Sanvito, S.: Magnetic interaction of Co ions near the (1010) ZnO surface. New J. Phys. **12**(083061), 14 (2010)
5. Ashtaputre, S.S., Deshpande, A., Marathe, S., Wankhede, M.E., Chimanpure, J., Pasricha, R., Urban, J., Haram, S.K., Gosavı, S.W., Kulkarni, S.K.: Synthesis and analysis of ZnO and CdSe nanoparticles. Indian Acad. Sci. **65**(4), 15–620 (2005)
6. Babu, B., Aswani, T., Thirumala Rao, G., Joyce Stella, R., Jayaraja, B., Ravikumar, R.: Room temperature ferromagnetism and optical properties of Cu2þ doped ZnO nanopowder by ultrasound assisted solid state reaction technique. J. Magn. Magn. Mater. **355**, 76–80 (2014)
7. Baolong, Y., Bingsuo, Z., Xiaochun, W., Guilan, Z., Guoqing, T., Wenju, C.: Preparation of ZnO nanometer powder and their ESR properties. Chin. J. Semicond. **16**(7), 558–560 (1995)
8. Behera, J.K.: Synthesis and Characterization of ZnO Nano Particles, M.Sc. Thesis, Department of Physics National Institute of Technology, India (2009)
9. Bieber, H., Versini, G., Barre, S., Loison, J.L., Schmerber, G., Ulhaq-Bouillet, C., Colis, S., Dinia, A.: Structural and magnetic study of hard-soft systems with ZnO barrier grown by pulsed laser deposition. Microelectron. J. **40**, 246–249 (2009)
10. Chen, Y.F., Wang, R.M., Zhang, H.Z., Sun, X.C., Zhang, Z.C., Xing, Y.J., Yu, D.P.: TEM investigations on ZnO nanobelts synthesized via a vapor phase growth, Micron **35**, 481–487 (2004)
11. Cong, C.J., Hong, J.H., Liu, Q.Y., Liao, L., Zhang, K.L.: *Synthesis*. Structure and ferromagnetic properties of Ni-doped ZnO nanoparticles, solid state communications **138**, 511–515 (2006)
12. Dadgostar, N., Investigations on Colloidal Synthesis of Copper Nanoparticles in a two-phase Liquid-Liquid System, M.sc. Thesis, University of Waterloo, Canada (2008)
13. Dakhloui, A., Jendoubi, M., Smiri, S.L., Kanaev, A., Jouini, N.: Synthesis, characterization and optical properties of ZnO nanoparticles with controlled size and morphology. J. Cryst. Growth **311**, 3989–3996 (2009)
14. Djaja, N.F., Saleh, R.: Composition dependence of structure and magnetic properties in manganese doped nanocrystalline ZnO particles prepared by co-precipitation. Mater. Sci. Appl. **3**, 245–252 (2012)
15. Djurisic, A.B., Choy, W.C.H., Roy, Y.A.L., Leung, Y.H., Kwong, C.Y., Cheah, K.W., Gundu Rao, T.K., Chan, W.K., Lui, H.F., Surya, C.: Defects in zinc oxide nano structures synthesis by a hydrotermal method. Adv. Funct. Mater. **14**, 856 (2004)
16. Duan, L., Zhao, X., Liu, J., Geng, W., Xie, H., Chen, S.: Structural, Thermal and magnetic investigations of heavily Mn-doped ZnO nanoparticles. J. Magn. Magn. Mater. **323**, 2374–2379 (2011)
17. Elilarassi, R., Chandrasekaran, G.: Synthesis, structural and magnetic characterization of Ni-doped ZnO diluted magnetic semiconductor. Am. J. Mater. Sci. **2**(1), 46–50 (2012)
18. Erdem, E.: Microwave power, temperature, atmospheric and light dependence of intrinsic defects in ZnO nanoparticles: A study of electron paramagnetic resonance (EPR) spectroscopy. J. Alloy. Compd. **605**, 34–44 (2014)
19. Fan, Z., Lu, J.G.: Zinc Oxide Nanostructures: Synthesis and Properties. Department of Electrical Engineering and Computer Science University of California, Irvine (2005)
20. Gao, X.P., Wanga, L.Z.: Nanoarchitectures of semiconducting and piezoelectric zinc oxide. J. Appl. Phys. **97**, 4, 044304, 1–7 (2005)
21. Ghobadi, N.: Band gap determination using absorption spectrum fitting procedure. Int. Nano Lett. **3**, 2 (2013)
22. Gluba, M.A., Friedrich, F., Lips, K., Nickel, N.H.: ESR investigations on hydrogen-induced hyperfine splitting features in ZnO. Superlattices Microstruct. **43**, 24–27 (2008)
23. Gubin, S.P., Koksharov, Y.A., Khomutov, G.B., Yurkov, G.Y.: Magnetic nanoparticles: preparation, structure and properties. Russian Chem. Rev. **74**(6), 489–520 (2005)
24. Gubin S.P.: Magnetic Nanoparticles, pp. 1–24. Wiley (2009)

25. Gupta, A., Bhatti, H.S., Kumar, D., Vermaa, N.K., Tandonb, R.P.: Nano and bulk crystals of ZnO: synthesis and characterization. Digest J. Nanomater. Biostruct. **1**(1), 1–9 (2006)
26. Hofmann, H.: Nanomaterials, Advanced Nanomaterials Course. Powder Technology Laboratory, IMX, EPFL, Version 1 (2009)
27. Hu, P., O'Neil, W., Hu, Q.: Synthesis of 10 nm Ag Nanoparticle Polymer Composite Pastes for Low Temperature Production of High Conductivity Films, vol. 257, issue 3, pp. 680–685. Centre for Industrial Photonics, Institute for Manufacturing, Department of Engineering, University of Cambridge (2010)
28. Hu, Y., Chen, H. J.: Preparation and characterization of nanocrystalline ZnO particles from a hydrothermal process. J. Nanopart Res. **10**, 401–407 (2008)
29. Ikeya, M.: New Applications of Electron Spin Resonance. Osaka University, Osaka (1993)
30. Ischenko, B.V., Polarz, S., Grote, D., Stavarache, V., Fink, K., Driess, M.: Zinc oxide nanoparticles with defect. Adv. Funct. Mater. **15**, 1945–1954 (2005)
31. Jayakumar, O.D., Salunke, H.G., Kadam, R.M., Mohapatra, M., Yaswant, G., Kulshreshtha, S.K.: Magnetism in Mn doped ZnO nanoparticles prepared by a co-precipitation method. Nanotechnology **17**, 1278–1285 (2006)
32. Kakazey, M., Vlasova, M., Juarez-Arellano, E.A., Torchynska, T.: EPR detection of sphalerite ZnO in mechanically treated ZnO + 0.1C nanosystem. Mater. Sci. Semicond. Process. **39**, 775–780 (2015)
33. Kappers, L.A., Gilliam, O.R., Evans, S.M., Halliburton, L.E., Giles, N.C.: EPR and optical study of oxygen and zinc vacancies in electron-irradiated ZnO. Nucl. Instr. Meth. Phys. Res. B **266**, 2953–2957 (2008)
34. Karamdel, J., Dee, C.F., Majlis, B.Y.: Effects of annealing conditions on the surface morphology and crystallinity of sputtered ZnO nano films. Sains Malaysiana **40**(3), 209–213 (2011)
35. Kodama, R.H.: Magnetic nanoparticles. J. Magn. Magn. Mater. **200**, 359–372 (1999)
36. Li, J.H., Shen, D.Z., Zhang, J.Y., Zhao, D.X., Li, B.S., Lu, Y.M., Liu, Y.C., Fan, X.V.: Magnetism origin of Mn-doped ZnO nanoclusters. J. Magn. Magn. Mater. **302**, 118–121 (2006)
37. Lia, j.H., Shena, D.Z., Zhanga, J.Y., Zhaoa, D.X., Lia, B.S., Lua, Y.M., Liuc, Y.C., Fan, X. W.: Magnetism origin of Mn-doped ZnO nanoclusters. J. Magn. Magn. Mater. **302**, 118–121 (2006)
38. Liu, H., Yang, J.I., Hua, Z., Zhang, Y., Yang, L., Xiao, L., Xie, Z.: The structure and magnetic properties of Cu-doped ZnO prepared by sol–gel method. Appl. Surf. Sci. **256**, 4162–4165 (2010)
39. Ma, X., Wang, Z.: The UV and blue light emission properties of Mn doped ZnO nanocrystals, China. Microelectric Eng. **88**, 3168–3171 (2011)
40. Mazhdi, M., Khani, P.H., Moghadam, M.C.: Effect of Mn low concentration on the optical properties of ZnO nanocrystals. Int. J. Nano Dim. **2**(2), 117–123 (2011)
41. Meulenkamp, E.A.: Synthesis and growth of ZnO nanoparticles. J. Phys. Chem. B **102**, 5566–5572 (1998)
42. Mhlongoa, G.H., Motaunga, D.E., Nkosib, S.S., Swartc, H.C., Malgasd, G.F., Hilliea, K.T., Mwakikunga, B.W.: Temperature-dependence on the structural, optical, and paramagnetic properties of ZnO nanostructures. Natl. Appl. Surf. Sci. **293**, 62–70 (2014)
43. Moballegh, A., Shahverdi, H.R., Aghababazadeh, R., Mirhabibi, A.R.: ZnO nanoparticles obtained by mechanochemical technique and the optical properties. Surf. Sci. **601**, 2850–2854 (2007)
44. Moontragoon, P., Pinitsoontorn, S., Thongbai, P.: Mn-doped ZnO nanoparticles: preparation, characterization and calculation of electronic and magnetic properties. Microelectron. Eng. **108**, 158–162 (2013)
45. Nirmala, M., Smitha, P., Anukaliani, A.: Optical and electrical properties of undoped and (Mn, Co) co-doped ZnO nanoparticles synthesized by DC thermal plasma method. Superlattices Microstruct. **50**, 563–571 (2011)

46. Noberg, N.S., Gamelin, D.R.: Influence of surface modification on the luminescence of colloidal ZnO nanocrystals. J. Phys. Chem. B **109**, 20810 (2005)
47. Nohavica, D., Gladkov, P.: ZnO nanoparticles and their applications-new achievements, 12–14. 10. Olomouc, Czech Republic, EU (2010)
48. Pashchanka, M., Hoffmann, R.C., Burghaus, O., Corzilius, B., Cherkashinin, G., Schneider, J. J.: Polycrystalline Zno and Mn-doped Zno nanorod arrays with variable dopant content via a template based synthesis from Zn(II) and Mn(II) schiff base type single source molecular precursors. Solid State Sci. **13**, 224–231 (2011)
49. Pearton, S.J., Norton, D.P., Heo, Y.W., Tien, L.C., Ivill, M.P., Li, Y., Kang, B.S., Ren, F., Kelly, J., Hebard, A.F.: ZnO spintronics and nanowire devices. J. Electron. Mater. **35**(5), 862–868 (2006)
50. Pearton, S.J., Norton, D.P., Ivill, M.P., Hebard, A.F.: ZnO Doped with transition metal ions. IEEE Trans. Electron Devices **54**, 5 (2007)
51. Pereira, A.S., Ankiewicz, A.O., Gehlhoff, W., Hoffman, A., Pereira, S., Trindade, T., Grundmann, M., Carmo, M.C., Sobolev, N.A.: Surface modification of Co-doped ZnO nanocrystals and its effects on the magnetic properties. J. Appl. Phys. **103**, 07D140–1–07D140-3 (2008)
52. Petrovic′, Z., Ristic′, M., Music′, S., Fabian, M.: Nano/microstructure and optical properties of ZnO particles precipitated from Zinc acetylacetonate. J. Mol. Struct. **1090**, 121–128 (2015)
53. Ramanujan, R.V.: Nanostructured electronic and magnetic materials. Sadhana **28**(1–2), 81–96 (2003)
54. Reddy, A.J., Kokila, M.K., Nagabhushana, H., Chakradhar, R.P.S., Shivakumara, C., Rao, J. L., Nagabhushana, B.M.: Structural, optical and EPR studies on ZnO: Cu nanopowders prepared via low temperature solution combustion synthesis, India. J. Alloy. Compd. **509**, 5349–5355 (2011)
55. Reddy, A.J., Kokilab, M.K., Nagabhushanac, H., Raod, J.l., Shivakumarae, C., Nagabhushanaf, B.M., Chakradharg, R.P.S.: Synthesis, luminescence properties and EPR investigation of hydrothermally derived uniform ZnO hexagonal rods. Spectrochim. Acta Part A: Mol. Biomol. Spectrosc. **139**, 262–270 (2015)
56. Sagar, R.V., Buddhudu, S.: Synthesis and magnetic behaviour of Mn: Zno nanocrystalline powders. Spectrochim. Acta, Part A **75**, 1218–1222 (2010)
57. Sahare, P.D., Kumar, V.: Optical and magnetic properties of Cu-doped ZnO nanoparticles. Int. J. Innovative Technol. Explor. Eng. **3**(6), 2278–3075 (2013)
58. Sati, P., Stepanova, A., Pashchenko, V.: Exchange broadening of EPR line in ZnO:Co. Low Temp. Phys. **33**(11), 927–930 (2007)
59. Schneider, J.J., Hoffman, R.C., Engstler, J., Dilfer, S., Klyszcz, A., Erdem, E., Jakes, P., Eichel, R.: Zinc oxide derived from single source precursor chemistry under chimie douce conditions: formation pathway, defect chemistry and possible applications in thin film printing. J. Mater. Chem. **19**, 1449–1457 (2009)
60. Sharma, V.K., Najim, M., Srivastava, A.K., Varma, G.D.: Structural and magnetic studies on transition metal (Mn, Co) doped ZnO nanoparticles. J. Magn. Magn. Mater. **324**, 683–689 (2012)
61. Song, C., Zeng, F., Geng, K.W., Wang, X.B., Shen, Y.X., Pan, F.: The magnetic properties of co-doped ZnO diluted magnetic insulator films prepared by direct current reactive magnetron co-sputtering. J. Magn. Magn. Mater. **309**, 25–30 (2007)
62. Sood, A.K., Wang, Z.L., Polla, D.L., Dhar, N.K., Manzur, T., Anwar, A.F.M., ZnO Nanostructures for Optoelectronic Applications, Optoelectronic Devices and Properties, 2011, Prof. Oleg Sergiyenko (Ed.), ISBN: 978-953-307-204-3
63. Stefan, M., Nistor, S.V. Ghica, D.: ZnS and ZnO semiconductor nanoparticles doped with Mn^{2+} ions. Size effects investigated by EPR spectroscopy. In: Kuncser, V., Miu, L. (eds.) Size Effects in Nanostructures (2014). doi:10.1007/978-3-662-44479-5_1
64. Thota, S., Dutta, T., Kumar, J.: On the sol–gel synthesis and thermal, structural, and magnetic studies of transition metal (Ni, Co, Mn) containing ZnO powders. J. Phys. Condens. Matter. **18**, 2473–2486 (2006)

65. Toloman, D., Mesaros, A., Popa, A., Raita, O., Silipas, T.D., Vasile, B.S., Pana, O., Giurgiu, L.M.: Evidence by EPR of ferromagnetic phase in Mn-doped ZnO nanoparticles annealed at different temperatures. J. Alloy. Compd. **551**, 502–507 (2013)
66. Tuyen, N.V., Canh, T.D., Long, N.N., Nghia, N.X., Trinh, B.N.Q., Shen, Z.: Synthesis of undoped and M-doped Zno (M=Co, Mn) nanopowder in water using microwave irradiation. J. Phys: Conf. Ser. **187**, 012020 (2009)
67. Vethanathan, S.J.K., Brightson, M., Sundar, S.M., Perumal, S.: Synthesis of Mn doped ZnO nanocrystals by solvothermal route and its characterization. Mater. Chem. Phys. **125**, 872–875 (2011)
68. Weil, J.A., Bolton, J.R.: Electron paramagnetic resonance: elementary theory and practical applications. Hoboken, Wiley Interscience (2007)
69. Yu, B., Zhu, C., Gan, F., Huang, Y.: Electron spin resonance properties of ZnO microcrystallites. Mater. Lett. **33**, 247–250 (1998)
70. Zhao, H., Zhang, Z., Zhao, Z., Yu, R., Wan, Y., Lan, M.: Preparation and characterization of novel spin labeled magnetic nanoparticles. Adv. Mat. Lett. **2**(3), 172–175 (2011)